Life, Death, and the Western Way of War

Life, Death, and the Western Way of War

LORENZO ZAMBERNARDI

OXFORD
UNIVERSITY PRESS

OXFORD
UNIVERSITY PRESS

Great Clarendon Street, Oxford, OX2 6DP,
United Kingdom

Oxford University Press is a department of the University of Oxford.
It furthers the University's objective of excellence in research, scholarship,
and education by publishing worldwide. Oxford is a registered trade mark of
Oxford University Press in the UK and in certain other countries

Published in the United States of America by Oxford University Press
198 Madison Avenue, New York, NY 10016, United States of America

British Library Cataloguing in Publication Data
Data available

Library of Congress Control Number: 2021953388

ISBN 978–0–19–285824–5

DOI: 10.1093/oso/9780192858245.001.0001

Printed and bound by
CPI Group (UK) Ltd, Croydon, CR0 4YY

In loving memory of my father, Livio Zambernardi (1943–2021)

Acknowledgements

The help of many people was essential to the writing of this book. I owe a special debt of gratitude to Michele Chiaruzzi, Asher Colombo, Mario Tesini, Konstantin Vössing, Srdjan Vucetic, and Alex Wendt, who read the entire manuscript and offered generous comments and remarks. Many others have contributed by advice and inspiration. Among those to whom I am particularly thankful are Filippo Andreatta, Nicola Antonetti, Giovanni Brizzi, Luigi Bonanate, Luciano Canfora, Emanuele Castelli, Ugo Fantasia, Ted Hopf, John Mueller, Geoffrey Parker, Arturo Carlo Quintavalle, Francesco Raschi, Dan Reiter, Randall Schweller, Matteo Truffelli, and Pascal Vennesson. I would like also to acknowledge the many exchanges with my fellow members of the research project *Death, Dying and Disposal in Italy: Attitudes, Behaviours, Beliefs, Rituals* (Prin 2015, 2015FR7MKM, Principal Investigator: Asher Colombo).

For constructive criticisms, which have helped me improve the manuscript, I am grateful to the three anonymous reviewers. At Oxford University Press I should like to thank Dominic Byatt for his support and encouragement and Céline Louasli and Hariharan Siva for their assistance in the production of the book. I thank the publishers for permission to revise and reuse material from the following articles: ‘Excavating Soldier Deaths: A Study of Changing Burial Practices’, *International Political Sociology*, 11/3 (2017) 292–307, and ‘Western Attitudes toward Soldiers’ Death: From the Early Modern Period to the Present’, *Mortality* (2020), DOI: 10.1080/13576275.2020.1864723.

I should also like to acknowledge all the participants at several venues to whom I presented the manuscript: the conference ‘Le regole della battaglia’ at the Fondazione Einaudi (Torino, December 2011); the conference ‘Nazione e nazionalismi’ at the Convento of Monteripido (Perugia, September 2016); the SGRI conference at the Fondazione Bruno Kessler (Trento, June 2016); and, finally, the seminar organized by the De Cive at the University of Parma (July 2021).

I am also deeply grateful to Ralph Nisbet, Lelio Pallini, and Hilary Walford. Their counsel and suggestions have done much to improve the literary quality of the book.

The most important thanks must go to Micòl and to my son, Andrea. Both gave me the necessary, peaceful distraction for writing a book on a not so pleasant topic such as death in war.

Oltretorrente (Parma), February 2022

Contents

Introduction

> Somebody has said that in order to know a community, one must observe the style of its funerals and know what manner of men they bury with most ceremony.
>
> Mark Twain, Buck Fanshaw's Funeral[1]

In his controversial *Medical Nemesis*, social thinker Ivan Illich (1976: 172) famously claimed that all 'disease is a socially created reality'. Without having to agree with such a contentious assertion, we might perhaps suggest that life and death are social constructions, in that the physical reality of dying has taken on a variety of meanings across time and societies (Walter 2012, 2020). Along with modifications in meaning, practices involving life and death have also changed, as on major ethical issues such as abortion, birth technology, suicide assistance, genetic engineering, and capital punishment. One of these changing practices regards how states fight in war. It is the main contention of this book that warfare is permeated by attitudes towards life and death, and the history and current reality of war are themselves the result of these changing attitudes.

The main goal of the present book is to trace when and how western soldiers—once regarded as simple *fighting tools*[2]—became the far less expendable beings that we know today. Or, to put it in Kant's terms, the study traces the process through which soldiers have been turned from mere military means into ends in themselves.[3] While today we might take it for granted that that the life of a soldier matters not only from a military point of view but also for political, humanitarian, and ethical reasons, that is nevertheless a major shift in how soldiers have been viewed throughout history. In fact, not only had political leaders and their civilian populations at home accepted for centuries that their combatants could blithely massacre one another at leisure, but they saw this brutal experience as natural, legitimate, and even appropriate (Luard 1986: 330; Mueller 1989: 38; Elias 2000: 163–5; Howard 2000: 1–2; Pinker 2012: 279–80).

[1] Twain (2008: 53).

[2] I paraphrase 'fighting tool' from Aristotle's definition of the slave as an 'animate tool'. See Aristotle's *Nicomachean Ethics* (1998: VIII, 1161^{b}5).

[3] I am referring to Kant's second formulation of the categorical imperative: 'For all rational beings stand under the *law* that each of them is to treat itself and all others *never merely as a means*, but always *at the same time as an end in itself*' (Kant 2012: 45).

Life, Death, and the Western Way of War. Lorenzo Zambernardi, Oxford University Press.
 DOI: 10.1093/oso/9780192858245.003.0001

The book argues that such a major transformation is largely the result of a shift in the social meaning ascribed to soldier deaths. It will be suggested that looking at death can somehow provide a privileged angle to understanding the value that societies attach to life. Not only is death the other side of the coin of life, but such a focus also highlights how *human material* is employed, wasted, and preserved on and off the battlefield. While it is almost impossible to demonstrate why societies in the past had different views of the value of human life (Laqueur 2015: 11), acknowledging that the meaning ascribed to death in war has changed over time, taking on a variety of forms and values, can throw light on the influence it has had on the development of modern warfare. In this respect, accounting for the changing meaning of military deaths amounts to revealing one of the central aspects of past and current warfare. Although tracing such a meaning is obviously far from explaining war in its entirety, it brings out one of its most significant aspects by highlighting how the individual is valued and used whether on or off the battlefield.

War, Death, and the Social Sciences

By looking at death as a social artefact whose meaning has historically changed, this enquiry intends partially to fill a void in International Relations (IR) theory by marking out a field of research that has so far received little attention. As Jessica Auchter (2015: 129) rightly noted, one of the greatest paradoxes of IR theory lies in the fact that its 'primary objects of study', 'namely, conflict and war, produce dead bodies en masse, yet' IR has failed 'to examine dead bodies in their complex potential'.[4] Except for a remarkable but relatively small scholarship on bodies in conflict, which focuses mainly on the victimization of civilians (Tirman 2011; Parashar 2013; Sylvester 2013; Gregory 2015, 2019; Wilcox 2015), scant attention has been paid to death in war. Though scholars of international politics have covered the issue of counting and explaining casualties in armed conflicts (e.g. Kalyvas 1999; Valentino 2004; Valentino et al. 2004; Lacina 2006; Lacina et al. 2006; Downes 2008; Lujala 2009), they have so far disregarded a fertile and promising area of enquiry: the changing *meaning* of military casualties.[5] Moreover, one problem with merely focusing on the

[4] For a similar argument, see Shah (2017a: 550–1).

[5] Notable exceptions are Wasinski (2008, 2011), Zehfuss (2009), Levy (2012), and Kaempf (2018). In *The Ultimate Experience* historian Yuval Harari (2008), the author of the international bestsellers *Sapiens* and *Homo Deus*, looks at the changing understanding of soldiers' battlefield experience. While there is some thematic overlapping between Harari's book and the present volume, the former focuses

war dead and their figures (that is, of transforming death into a mere statistic) is that it overlooks the practices, ideas, and norms that made those dead possible in the first place.

Despite the centrality of death in the matter of war, even military analysts and historians seem to have failed to give death a central, prominent place in their analysis. They have studied important topics such as the link between weapons lethality and casualties (Adamson 1977; Dupuy 1984), brutality, carnage, and ruthlessness on the battlefield (Ehrenreich 1997; Hanson 2001a; Hull 2005; Scheck 2005; Kramer 2007), as well as the relationship between combat and culture (Lynn 2003), but even the body of scholarship that looks at how killing has changed on the battlefield (Price 1997; Bourke 1999; Thomas 2001; Slim 2007; Stephenson 2012; Shah 2017b) has not made death central to the understanding of war.

By interpreting war as an extension of society, sociologists and social historians have greatly contributed to showing that warfare is more than technological, tactical, and strategic processes.[6] Their basic proposition, shared by this research, is that, in order to understand war, it is necessary to keep society and armed conflict within the same analytical framework. However, sociological works on war have been mainly interested in the generative and transformative power of war over social relations (Tilly 1992; Kestnbaum 2005; Malešević 2010a; Mann 2012a, b; Wimmer 2014; Scheidel 2018).[7] In this research a somewhat different view is proposed. The book explores how changing notions of the subject (that is, the soldier) at the social level have shaped the conduct of war.

The Emergence of the Individual

The narrative emerging from the empirical evidence will show that the story of attitudes towards soldier deaths is the story of a gradual, increasing process of individualization in the social meaning attached to human loss in war. Such a development, which took centuries to emerge in full, was neither simple nor linear. It was a process that the state was temporarily able to frame in the collective narrative of the nation, but that has ultimately seen the increasing

on the meaning that soldiers themselves ascribed to the encounter of death in battle; this research, by contrast, concerns mainly the meaning attached to death in war by society.

[6] For an assessment of the contribution of the classics of social thought to the study of war and violence, see Malešević (2010b). See Hutchinson (2017) for a study that connects warfare and war commemoration with the rise of nationalism.

[7] For an exception, see McSorley (2014).

importance of the life of the individual soldier. In particular, a long age in which military deaths were totally unproblematic and accepted as an ordinary outcome was followed by a shorter period when death in war was regarded as a glorious, holy sacrifice for the nation, which culminated in a sensitivity to friendly casualties that is typical of the era in which we live. While commoners in what is generally referred to as the early modern period were seen as anonymous masses with no specific individual identities, since the mid-1800s they have increasingly attained individual and equal status. For about a century, however, nationalism and patriotic sentiment were able to channel the rise of soldier individuality into the holistic narrative of the nation state, thus justifying the mass sacrifice of combatants. But in the decades following the Second World War soldier individuality fully emerged, markedly affecting the possibility of treating soldiers as mere military means. I shall refer to these epochal shifts as a transition from *bare death* to *sacrificial death* and, finally, to *irrecoverable death*.[8] The overall outcome of this process is a reversal in the meaning ascribed to death in war: from something that appeared as part of the natural order of things to an intolerable loss, which must be avoided or at least minimized on the battlefield.

The three stages do not mark a linear development and are to some extent arbitrary. For one thing, historical continuity invites questions about the appropriateness of dividing history into different ages and periods—and, indeed, some overlaps are to be found among the three periods. Moreover, perfect synchrony in the pace of change among different national cases does not exist, nor is there any exact coincidence within geographical areas. Each epoch shows considerable change—and some continuity—over the centuries of warfare covered by this research. Nevertheless, I would argue that the three different periods are dominated by a single unifying meaning attaching to soldier deaths. While death in the first historical period is a meaningless cessation, what I call *bare death*, in the second and third epochs it becomes highly significant for societies and states.[9] However, though attitudes towards soldier deaths in the third stage are as socially meaningful as in the second one, the two meanings are in complete opposition: while the meaning of military deaths in the second stage of development justifies mass slaughter—indeed the

[8] These three different historical meanings of death in war should be understood as Weberian ideal-type orientations to death. As with any ideal type, no perfect correspondence with empirical reality can be expected: elements of all three meanings are to be found in every epoch. However, as Weber (1949: 90–107) taught us, ideal types are useful, not if they can be completely recognized in the empirical world, but rather when they are able to clarify and create order in the complexity and richness of empirical reality.

[9] For a different usage of the concept of 'bare death', see Sherry (2018).

term that describes such an attitude here is *sacrificial death*—the meaning of the third and latest stage signifies care and protection for one's own combatants, what may be called *irrecoverable death*. Thus, the dominant change in the meaning of soldier deaths has been from insignificance to a sacrificial attitude and, finally, to aversion.

In tracing the process through which soldiers have been turned from an amorphous collectivity into distinct individuals, this research also aims to shed light on the origins of a variety of present phenomena such as the 'differential allocation' of life and death in war (Butler 2004, 2009; Zehfuss 2009), 'post-heroic warfare' (Luttwak 1995), 'risk-transfer militarism' (Shaw 2005; Kaempf 2018), and all those military practices that are meant to minimize friendly casualties (T.W. Smith 2008). The analysis, moreover, not only advances knowledge in an understudied area of enquiry, but also opens several avenues for future research, especially in a comparative perspective with those cultures that significantly diverge from the developments described in this book.

This leads us to the second purpose of the research, which concerns the relationship between military means and political ends, in particular how modifications in social conceptions of the soldier (that is, one of the main military means throughout history) have affected the utility of using force in the current epoch. In order to clarify this second goal of the book, a few words on the origins of the changing meaning of soldier deaths are in order.

The Rise of the Individual and its Consequences for Western Warfare

The process of individualization of soldiers' lives and deaths described in the pages that follow is, clearly, a complicated process resulting from multiple interrelated and unrelated causes. But, whatever its economic, religious (Le Goff 1984; Brown 2003, 263–4; Siedentop 2015), intellectual (Batkin 1992; Burckardt 1944), and political roots (Ullmann 1967: 62–98; Macfarlane 1978; Stourzh 1996: 314)—a question beyond our reach—it is not too precipitate to argue that it was bourgeois society, armed with its liberal political ideology, that spread throughout the western world the revolutionary conception of the individual as a unique being with a particular, distinct identity, and specific personal rights.[10] By destroying the remnants of the *society of orders* with its

[10] A considerable number of historians, philosophers, and literary scholars have traced the roots of the modern conception of the individual to the Renaissance (Burckhardt 1944; Garin 1975; Batkin

hierarchical notions about human nature and, more generally, social status as a determinant of a person's role in a specific community, bourgeois society and liberal ideology strongly contributed to the process of moral and legal equalization of humans as against various types of status difference and discrimination that characterized medieval and early modern Europe (Rosanvallon 2011). In so doing, increasing awareness of the supreme worth of the person has produced a *society of individuals* (Beck 1992; Dumont 1992; Elias 2001a) in which the meaning of soldiers' lives has changed radically as well.[11] Stated otherwise, the fact that the individual has become the 'organizing social role in the West' (Siedentop 2015: 2) has also slowly but inevitably underpinned the views that modern societies have of their combatants.

It goes without saying that casting soldiers as, and *qua*, individuals is a by-product of this process. Today, economic prosperity, constitutional government, the proliferation of democracies around the world, and the protection and promotion of human rights are generally seen as major achievements of the rise of modern individuality (Benhabib 2006; McCloskey 2016; Pinker 2018). Concerning the field of war, an emerging bulk of scholarship has explicitly discussed the notion of the 'individualization of war', pointing out how the increased prominence of the individual in the theory and practice of war is challenging the primacy of the state as the main referent, responsible agent, and provider of security (McMahan 2009; Strawser and McMahan 2013; Voelz 2018; Welsh 2019).

Some scholars have also argued that democratic societies, formed of free men and women who willingly fight for their nations, are more effective in waging war than those authoritarian regimes that deny basic individual and political rights (Hanson 2001b; Reiter and Stam 2002: 58–79).[12] Contrary to such scholarship, this study is not meant to provide a progressive account of the development and spread of the liberal understanding of the individual.

1992; Chabod 2008; Biow 2015) and even to the late medieval period (Morris 1972; Greenblatt 1980; Le Goff 1984; Brown 2003; Siedentop 2015). While many studies have convincingly shown that the modern conception of the individual began to emerge in these historical periods, bourgeois society, with its emphasis on critical thinking, freedom of thought, and human agency, is the place where not perhaps the origin but at least the diffusion of this nascent reality is to be located. This is why I employ the term 'spread' in reference to the contribution of the bourgeois ethos.

[11] See also Strathern (2005), who speaks of an 'age of individualism'.

[12] This is an argument that partly recalls Machiavelli's idea about the superiority of civic militias over mercenaries, an argument developed in a variety of writings including *The Prince* (e.g., Machiavelli 1998: ch. 12, 'How Many Kinds of Militaries There Are'). As secretary of the Florentine Republic, Machiavelli tried to turn this idea into reality with the creation of popular troops (*Ordinanza*). However, the humiliating sack of Prato (1512) by the Spanish army, with fleeing soldiers who failed to stand the ground during a two-days siege, showed the limits of civic militias against a superior number of better trained, equipped, and motivated combatants.

Although it is true that principles such as 'legal equality' and 'personal and human rights' are some of the greatest achievements of bourgeois societies, these pages will focus on how the emphasis on the primacy of the individual has further eroded the effectiveness of western warfare as an instrument of foreign policy. In particular, the modern, liberal conception of the soldier has had the unintended consequence of jeopardizing the Clausewitzian relationship between military means and political ends.

As shown by recent events, the most advanced arsenal in the history of war has failed to defeat relatively small insurgencies in places such as Iraq and Afghanistan, where western armed forces were unable to impose their will on incomparably weaker enemies. Even a cursory look at recent military interventions makes it patently clear that every major western operation in the last few decades—from Iraq and Afghanistan to Libya—has far from achieved its political objectives. How come western warfare, based on the best modern technology available and backed by some of the wealthiest economies in the world, has resulted in a systematic series of military failures? Of course, there is no single reason or variable (to use social-science jargon) that can account for these military outcomes. Not only has victory many fathers, as President Kennedy ironically and famously put it, but so has defeat; the results of armed confrontations cannot be reduced to simple explanations based on one or a few factors. War is so complex as to make what Hans Morgenthau (1965: 92–3) labelled the 'method of the single cause' not only mistaken from an epistemological viewpoint,[13] but also dangerous when it is embraced as a doctrine for the conduct of foreign policy and war (Zambernardi 2016b). More modestly, I here intend to argue that current attitudes towards soldier deaths have further eroded the utility of military power, especially in those irregular conflicts in which western states mostly engage. While in the past the greatest threats to national security were posed mainly by conventional enemies, today's security challenges also come from adversaries that fight irregularly and asymmetrically. In particular, the capital-intensive strategy Caverley (2014) employed by western states, which relies heavily on firepower, contradicts the nature of the irregular conflicts in which they are mostly involved. A high level of firepower and mechanization is inappropriate in an age when the nature of war demands thorough control over the territory and dispersion in the enemy population. Paradoxically, immense military might has made western countries weaker rather than stronger, ineffective rather than successful.

[13] On Morgenthau's critique of the 'method of the single cause', see Zambernardi (2020a).

As I will argue in the final chapter, the predominance of nonconventional conflicts in the present era would recommend the increasing use of the foot soldier or, to borrow Carl Schmitt's terminology (2004), of those combatants with a 'telluric character'.[14] Yet western states are still trying systematically to remove soldiers from the ground. Such a preference for employing machines rather than soldiers, even when faced with the threat of military failure, is neither coincidental nor a product of ignorance. It results directly from the main subject matter of this research: that is to say, from long-term changes concerning attitudes toward soldier deaths in western societies.[15] Hence, the book sets out to illustrate both why death-dealing technology, such as ballistic missiles, drones, and air power more generally, is not a good alternative to boots on the ground, and why soldiers cannot be employed as they were in the past. Thus, not only is tracing the origins and development of soldiers' individuality an important chapter in modern and contemporary history, but it can also lend us an interesting perspective on the nature of current warfare and on some of the limits of coercive force in the present era.

The remainder of this introductory chapter will serve, respectively, to clarify the research puzzle from which this study springs (that is, to show that a major change has occurred in how military deaths are viewed by western societies), to define the time frame of the enquiry and to discuss the method employed, and, finally, to describe the organization of the volume.

[14] At the end of the eighteenth century the military importance of the foot soldier began to diminish. With the use of mobile battlefield artillery, the cohesiveness of massed infantry was disrupted. In the middle of the nineteenth and early twentieth centuries, the battle-winning capacity of infantry was further reduced by the introduction of new, long-range, more accurate firearms, which could also multiply the army's firepower. 'By the end of the first decade of the twentieth century', wrote John Keegan and Richard Holmes (1985: 14), 'the era of the primacy of man in warfare was drawing to a close' and 'the era of the primacy of the machine was at hand'. Furthermore, in the course of the twentieth century, by promising victory at a tolerable human cost, the mechanization of war and the technological factor continued to increase.

[15] Jason Lyall and Isaiah Wilson III (2009) have recently argued that the key variable explaining states' increasing incapacity to defeat insurgencies in the twentieth century is the growing mechanization of their armed forces. In particular, according to Lyall and Wilson III (2009: 79–80), governments could not opt for more 'primitive' and efficient types of warfare because of a 'shared set of norms that govern states' beliefs about what constitutes the most appropriate means of military organization … Mechanization was, in some senses, a ticket to international legitimacy'. Although this research broadly agrees with Lyall and Wilson's claim that the increasing mechanization of western warfare is relevant in explaining states' growing inability to defeat insurgencies, their explanation raises two problematic points. First, the authors do not provide an actual analysis of the international norm suggesting that mechanization is the most appropriate means of military organization: they simply discuss it in the rather small space of two pages. Secondly, it is noteworthy that several norms were broken or disregarded either when victory in war was at stake or after a burning defeat. With its focus on changing attitudes towards soldier deaths, the present study will try to account also for the puzzle of over-reliance on mechanized warfare against insurgencies.

Changing Attitudes towards Soldier Deaths?

'Have there been times and places where casualties on one's side have been viewed as a benefit rather than a cost?,' ask Peter Feaver and Charles Miller (2014: 146). The answer to this question, as they suggest, is almost certainly negative. However, for most of human history, preoccupation with the lives of combatants primarily lay in the military performance they could offer on the battlefield, since more troops meant more fighting power.[16] Indeed, in the past, when the exigencies of war so required, combatants' lives were highly expendable. This attitude, whereby soldiers' lives were valued merely for their military contribution, appears to differ significantly from the present position, in which protecting the lives of combatants seems to be a political, social, and moral requirement for their deployment in the theatre of war. The point is that today individual lives are deemed valuable beyond soldiers' fighting capacities, whereas soldiers in the past were employed with caution only for the safety of the army. Although death has always been an ordinary occurrence in war, it has now become a burning social and political issue, which is met with criticism and in some instances even with resistance.

According to John Mueller (1973), sensitivity to casualties for the United States can be traced back to the wars in Korea and Vietnam. In particular, he argued that public support in those conflicts was negatively related to the log of casualties. While the Second World War was extremely popular in the United States despite high levels of war dead, according to Mueller the conflicts in Korea and, especially, Vietnam were characterized by a decline in support as casualties soared.

Mueller's view has been challenged by a variety of studies suggesting that the relationship between casualties and public opinion is more complex than that. In particular, people's sensitivity to casualties is viewed as an outcome of the public perception of costs and benefits based on a series of factors such as the framing of the war's costs (Boettcher and Cobb 2009), geographical proximity to casualties (Althaus et al. 2012), the degree of elite consensus (Burk 1999), the importance of the issues at stake (Larson and Savych 2005), and patterns of elite conflict (Berinsky 2009). According to Gelpi et al. (2009) the key variable accounting for public support of war is the perceived likelihood of success in war. In particular, these scholars have criticized Mueller's thesis by arguing

[16] What Stephen Biddle (2004: 2–3) defines as 'modern system force employment' (i.e., the systematic reduction of troops' exposure to highly lethal enemy fire) seems to fall into this category, since it is meant to 'perform meaningful military missions' rather than protect a soldier's life as an end in itself.

that Americans are more 'defeat-phobic' than 'casualty-phobic', meaning that American society is concerned more about losing a war than about losing soldiers. In other words, the American public makes rational judgements about the potential costs (largely in terms of casualties) and benefits of a particular intervention.

Although most of these refinements are important in the current debate on the relationship between war and public opinion, they do not fundamentally challenge the idea that casualties have become a sensitive issue for the American public and for European societies too (Everts and Isernia 2015). Indeed, none of these studies denies that western publics are now sensitive to friendly casualties. The point is that American and European societies and their political and military elites used to be largely unmoved by casualties. By contrast, victory is now measured not only in terms of battles won but also in terms of lives lost. That does not imply that a handful of friendly casualties are considered intolerable, but it does mean that soldiers' lives have become a paramount and unprecedented concern. While western societies are far from being casualty phobic and the degree of sensitivity to friendly casualties may vary from country to country and even among different sections of the same society (Kriner and Shen 2010), what seems indisputable is that there has been a considerable change in attitudes towards combat casualties. Even critics of the idea of a casualty-phobic public, such as Gelpi et al. (2009: 12, 244), admit that a radical transformation has taken place: 'Whereas in previous wars, casualty concerns arose after tens or even hundreds of thousands of fatalities', they argue, 'now the concerns are arising even when the death toll is only in the tens or hundreds'. The public, they conclude, 'has indeed become more casualty sensitive'. Thus, regardless of the strict relationship between casualties and public support for war, it seems unquestionable that western states and publics see the avoidance of casualties as a major priority in military interventions abroad.

A *Longue Durée* View on the Changing Meaning of Soldier Deaths

As already mentioned, this study offers an empirical account of the change in the social meaning of soldier deaths. The research adopts the *longue durée* mode of enquiry, emphasizing the value of a cross-disciplinary, long-range vision of the past for understanding complex contemporary phenomena.[17]

[17] On the importance of the *longue durée* approach, see Braudel (1958) and Guldi and Armitage (2014).

Such a perspective allows us to go beyond the boundaries of national narratives and to outline the rise and development of the process that has turned the soldier from a fighting tool into a precious individual.

Like many *longue durée* analyses, this is largely an interdisciplinary work of synthesis, combining together scholarship and historical evidence from a variety of disciplines. In particular, the research moves back and forth between sociology and anthropology and between cultural and military history; it is largely an interdisciplinary work pooling research and evidence from fields that have not been able to communicate adequately. However, the study goes beyond mere synthesis of adjoining academic literatures. Its actual worth lies in merging the historical evidence from different fields to produce a novel narrative on a theme that has so far been largely neglected.

The story told in this book begins from early modern Europe, because that period marked both the transition from the age of the warrior to the age of the soldier and the decline of armoured knights in European warfare. The age of knights began to fade in the sixteenth century and was replaced by what the Italian diplomat and poet Fulvio Testi called 'the century of the soldier' (in Parker 2005b: 149).[18] The primary, though not exclusive, focus on foot soldiers in the present study results mainly from a simple truth: historically, when employed, they were the combatants most likely to die in war.

Chronologically, this is certainly an ambitious enquiry, especially in the light of a complex and diverse subject matter, which sacrifices depth to examining a long historical evolution in a wide geographical context. The risks of this approach are obvious: omission, problems concerning the selection of evidence, consistency of the evidence itself, impressionistic use of the empirical material as a cultural and social indicator, and the like. However, I feel that the ambition of tracing and understanding the evolution of the different meanings

[18] There is a long and interesting debate as to the beginning of early modern history. Most scholars trace its origins to the last decade of the fifteenth century, because of the discovery of America in 1492 and, more interestingly for us, the military expedition in Italy by Charles VIII in 1494. The so-called Italian wars, a series of military campaigns that took place from 1494 to 1559, revealed the diminishing role of heavy cavalry in comparison to infantry, which became the decisive combat arm on European battlefields. The Italian wars witnessed both the last appearance of massed armoured cavalry at the battle of Pavia in 1525, and the systematic employment of massed infantry. Given the vulnerability of fully armoured mounted warriors to firearms and solid infantry formations of pikemen, heavy cavalry—the major killing arm in medieval warfare—became only one element of the three arms of early modern European warfare (Showalter and Astore 2007: pp. xii, 71). With such a military change, which was reflected in the army that Charles VIII employed in the Italian Peninsula, 'the wars of the knights were over' (Howard 1976: 19). However, it should be noted that cavalry proved decisive on a variety of occasions, such as Balaclava during the Crimean War, where the Heavy Brigade (not to be mistaken with the Light Brigade) proved successful against the Russians. One of the latest recorded instances of successful cavalry charges occurred at Isbuscenskij (near the junction of the Don and Khopyor rivers) on 24 August 1942, when the Italian cavalry regiment *Savoia Cavalleria*, armed mainly with sabres and hand grenades, attacked a Soviet infantry regiment armed with machine guns and mortars.

attached to military deaths justifies my lack of shyness in embracing a span of several centuries. Although neat generalizations for such a long historical period might appear to blur important distinctions and overlook significant phenomena, there is no doubt that—as the eminent historian of death Philippe Ariès (1981: p. xvii) contended—'the errors [the researcher] will not be able to avoid are less serious than the anachronisms to which he would be exposed by too short a chronology'.

The whole period is not meant to be surveyed with equal attention, though. Rather than aspire to any systematic account, the present analysis focuses mainly on the salient stages where a certain practice emerged. For obvious reasons of space, telling the whole story is hardly possible in one study, as it would entail a monumental, overwhelming effort for a single researcher. Thus, differences in emphasis and discussion of particular historical epochs, conflicts, and societies (for example, the greater length of Chapter 3 compared to Chapters 2 and 4) reflect their relevance to the phenomena at issue.

A few remarks on the theoretical approach and method are now in order. The procedure employed to capture the changing meaning of death is admittedly somewhat old-fashioned. There is neither quantitative nor in-depth qualitative analysis. No experiment, no regression, no thick description, no process tracing, no multi-method approach, no genealogy, and no other recent discoveries of social-science methodology are employed here. However, in order to avoid both the typical eclecticism of some cultural and social history (Winter 1995: 10) and the charge of resorting excessively to the anecdotal approach, as, for example, the work of Ariès has been accused of (Stone 1981: 251), the enquiry is theoretically underpinned by Ernst Kantorowicz's concept (1997) of 'the king's two bodies' and rigorously organized around three central types of evidence that can help reconstruct the changing meaning of soldier deaths: employment of soldiers in war, disposal and burial practices, and war memory. These three practices show how combatants are used or not used on the battlefield when they are alive, how soldiers' corpses are disposed of, and how the war dead are remembered and commemorated after their death. Thus, the investigation is concerned not with all details of soldiers' lives, but only with those military, social, and ideational elements that disclose something interesting about how their life and death were valued and appreciated; with those elements, in other words, that reveal soldiers' place 'in the life of a society' (Corvisier 1985: 368).

Before I describe how the volume is organized, two further points are worth mentioning. First, though there are remarkable and interesting variations in the attitudes towards death across the cultures, the emphasis is here placed on historical variation in one single area, namely western Europe and North America. Taking such a perspective is meant neither to reject the significance of other cultural and geographical areas nor to show the uniqueness of this experience. I confess I adopted the expression *western world* with some hesitation, given the ambiguity and even the indeterminateness of the notion (Appiah 2016), but will nonetheless employ it for two main reasons. On the one hand, as far as the rise of soldiers' individualization is concerned, what we have come to call the West appears to show profound cultural homogeneity. On the other, this narrow perspective is justified by the fact that it is only fair for the author to declare his shortcomings openly, as other scholars will certainly have much to add.

Secondly, while what has been said so far is not concerned with questions about paradigms, research traditions, or schools—these issues will not be discussed in the book—it is necessary to clarify how the topic is approached. The simple argument developed in the research is that people's understandings of the individual (or its different social constructions) have shaped how western states fight. Accordingly, this is intended as a contribution to the growing body of scholarship that grants considerable importance to cultural facts in the investigation of political and military phenomena. In so doing, the study forms part of an expanding corpus of literature that sees norms and culture as variables essential to understanding international politics (Katzenstein 1996; Lebow 2008; Vucetic 2011) and war (Legro 1995; Kier 1997; Hanson 2001a; Thomas 2001; Lynn 2003).[19] In particular, the present analysis belongs to that stream of research that, in the tradition of Michael Howard (1976), attempts to keep society, armed forces, and war within the same analytical framework.

One final point must be emphasized. In the present book I do not argue that the meaning ascribed to military deaths is the most important dimension of war. There is no doubt, for example, that today it is modern technology that makes it possible to wage war from a safe distance. Likewise, there is no doubt that modern media offer an unprecedented window onto conflicts where soldiers are deployed, providing shocking and touching images of combat, violence, and corpses. Thus, despite the importance of the social construction of meaning, we should keep in mind what one of the leading constructivist scholars wrote about the role of brute facts for the social world: the 'composition of material capabilities at any given moment helps define the possibilities of our action' (Wendt 1999: 113). What I try to show in this

[19] Unlike some important studies in the academic literature, however, this research is not concerned to show that ideational and cultural variables have more explanatory power than material ones.

research is, by contrast, that a crucial cultural and social dimension of warfare has been largely neglected so far, which is well worth considering because of its extraordinary interest.

Structure of the Book

The volume is arranged as follows. Chapter 1 aims to set the present enquiry within the framework of the existing literature concerning death and, in particular, death in war. Some theoretical discussion will also be provided of the 'multiple bodies of soldiers' along with a few suggestions on how to reconstruct the changing meaning of soldier deaths from the study of combat, body disposal, and war memory. This part will attempt to offer a few theoretical insights into the ensuing historical analysis. The chapter concludes by providing a unified account of what brought about the changes outlined in the volume and a brief evaluation of the existing alternative explanations.

Chapter 2, 'Bare Death', investigates the meaning of soldier deaths in early modern Europe. This first episode of the story, up to the middle of the nineteenth century, is relatively uncomplicated. As it will be argued that soldier deaths were considered as a natural occurrence that raised no political or ethical sentiment—an outlook clearly visible from disposal and commemorative functions—the chapter also describes how states used soldiers on the battlefield. The main military tactics of the epoch will be outlined, and it will be contended that the main motivation of battlefield tactics was to overpower enemy forces, regardless of human cost.[20] Considerations over casualties certainly played a role in warfare, but solely in terms of combat power. This is the case even with the most important tactical innovation of the time, the passage from squares to lines. At the end of the chapter, in the section entitled 'Intermezzo', I briefly reconstruct the emergence of a radical change in societal attitudes to death that developed in the late eighteenth century and in the first half of the nineteenth century. Such a change, I suggest, lay at the root of the changing attitudes to soldier deaths that would materialize in the following decades and that are clearly visible in disposal practices and war commemoration.

In Chapter 3, 'Sacrificial Death', the changes that took place between the mid-nineteenth century and the first half of the twentieth century will be

[20] Following Clausewitz (1984: 128), I refer to tactics as the doctrine of the use of armed forces in battle and to strategy as the doctrine of the use of individual battles for the purpose of the war.

discussed to argue that a radical alteration took place in the meaning that western societies gave to military casualties. In the second half of the nineteenth century, western states started to celebrate the ordinary war dead as national heroes whose sacrifice needed to be publicly recognized and commemorated. It was in that period that the death of soldiers ceased to be socially meaningless and started to have public importance entailing monuments and other forms of commemoration. Thus during and after the American Civil War, the Franco-Prussian conflict, and the First World War a very cult of the war dead can be observed, creating in turn a form of social significance for the death of ordinary soldiers that was totally absent in the early modern era. As we shall see, in this period the dead body of the soldier is invested with a value that cannot be deduced from its corporeality. The term *sacrificial* is, indeed, used in the sense of human lives being invested with a sentimental, political, and almost religious meaning. The transformation of soldiers from fighting commodities to human beings had begun. As shown further on, the sacralization of soldier deaths was a transitional phase between indifference towards soldiers and the typical casualty sensitivity of our current epoch.

Chapter 4, 'Irrecoverable Death', focuses on the second and latest radical change in the social meaning of soldier deaths, which emerged fully after the Second World War. In that period the glorification of death in war lost its social attractiveness and turned into the aversion to death typical of the present epoch. While soldier deaths maintained their social importance, their meaning was gradually reversed from glorification to aversion. While soldiers in the previous period were granted the privilege of memory in death, now the privilege also involves the protection of their lives on the battlefield. As a result, death has lost its sacrificial connotation, and the preservation of soldiers' lives has become an end in itself. With such a change, soldiers ceased to be considered as an anonymous mass and started to be conceived as persons with a right to life even in that 'marketplace dedicated to the exchange of casualties' (Keegan and Holmes 1985: 31) that we call war. It will be shown that such a metamorphosis has produced military strategies and tactics devised to minimize military casualties, such as the preference for bombing from a safe distance (for example, air power, ballistic missiles, unmanned weapons, etc.), the increasing use of indigenous forces, and, when land warfare is inevitable, the adoption of *force protection* measures. The discussion will also indicate that these military tactics and practices serve a specific function: making war possible for states that are highly sensitive to military fatalities. For no western society, no politician, officer, or soldier, would today accept the casualties of the past.

Finally, Chapter 5, 'Epilogue', is intended to show that sensitivity to military deaths does not merely shape how military force is exercised, but has also diminished the utility of war as an instrument of foreign policy. In particular, it will be suggested that the values western societies currently attach to individual soldiers are jeopardizing the Clausewitzian relationship between military means and political ends, thus endangering the attainment of the political goals that leaders assign to their wars, especially in the most common type of conflict in the present era: irregular wars. In other words, the full rise of the individual in the contemporary western world has reversed the traditional relationship between the soldier and the state and, in so doing, further eroded the utility of military power.

Since providing a full description of this diverse and complex subject matter is impracticable, the present analysis cannot avoid being tiny in proportion to the vastness of the area. Admittedly, the subject involves several disciplines and, as such, is filled with problems of method and sources. Limited though it may be, the study nevertheless attempts to offer a novel perspective from which to view and understand the effects of the familiar phenomenon of death in war.

1

Combat, Body Disposal, and War Memory

According to Thomas Hobbes (2012), the fear of death lies at the root of political order. In particular, the possibility to escape the state of nature and form a commonwealth stems from individuals' concern for self-preservation. While the great English philosopher posits death at the centre of political modernity, the eminent anthropologist Ernest Becker (1973: p. ix) went even further, putting death at the heart of human existence: in his Pulitzer-prize winning book *The Denial of Death*, he argued that 'the idea of death' is 'a mainspring of human activity—activity designed largely to avoid the fatality of death'. The 'fear of it', Becker famously claimed, 'haunts the human animal like nothing else'.[1]

Doubtless, as a final separation from what we value, death evokes fear, revulsion, and terror. Despite the exceptional know-how provided by modern medicine, at the moment of death we still appear as powerless as we were in the past. However, while death is universal and the fear of dying might appear natural and innate at a personal level, different cultures have ascribed to death a variety of meanings, images, emotions, and values. Not only have archaeologists (Parker Pearson 2000), social anthropologists (Metcalf and Huntington 1991; Lomnitz 2005; Laqueur 2015), and historians (Stannard 1977; Vovelle 1983; Ariès 2009) revealed a high degree of diversity in the ways societies have confronted the problem of mortality, but throughout history different societies have developed opposing conceptions of death: some saw it as a natural and unproblematic phenomenon, some despised and demonized it, whereas others beautified and deified it. Even today, death may be considered by some individuals as a friendly companion and by others as a terrifying enemy (Kamm 2020). In other words, death has not always been perceived and understood in one and the same way; its meaning has not been written in stone but rather has changed over time. In the twentieth century, for example, western societies appeared to have moved from the idea of a 'good death' as an accepted social

[1] Ernest Becker is the main source of inspiration for *terror management theory* (TMT). TMT suggests that awareness and fear of mortality produces terror, which human beings manage by embracing specific cultural values and world views functionally meant to produce self-esteem and a sense of existential security. For an application of the theory to American society in the wake of the 9/11 terrorist attacks, see Pyszczynski et al. (2003). For a well-known application of TMT to international politics, see Mitzen (2006).

Life, Death, and the Western Way of War. Lorenzo Zambernardi, Oxford University Press.
 DOI: 10.1093/oso/9780192858245.003.0002

rite of passage towards a sequestered biomedical practice in healthcare facilities where death is sanitized and prolonged. Familiarity with the occurrence of death, which used to make people aware of their own final end, has now turned into distancing from death in the lives of most individuals (Ariès 1981; Giddens 1991; Elias 2001b; Kellehear 2007, Sozzi 2009).

From this viewpoint, death is more than a simple brute physical fact marking the end of human life, which does not imply—of course—that life and death have no material and biological constitution,[2] but rather indicates that the meanings and values that societies have attached to death and to different *lives* have changed historically (Kearl 1989; Zelizer 1994; Ariès 1996; McMahan 2002). Death is both a biological and a social phenomenon, and one need not be a postmodern theorist to admit that its different meanings are socially constructed.

Based on the idea of death as a changing social fact and as an engine of social phenomena, the present chapter divides into three main parts. First, relying on Michel Vovelle's analysis, the significance of death as a cultural artefact will be briefly illustrated. Secondly, drawing on Ernest Kantorowicz's concept of the 'king's two bodies', the chapter will discuss how the meaning of soldier deaths can be pieced together through a systematic study of combat, body disposal, and commemorative practices. The chapter concludes by critically evaluating the existing accounts concerning changes in attitude towards military deaths and by sketching an alternative explanation.

Studying Death as a Cultural Artefact

In his monumental study *La Mort et l'Occident de 1300 à nous jours*, the eminent French historian Michel Vovelle (1983) argued that there are three main macro-approaches to the study of death: *la mort subie* (death as a brute fact), *la mort vécue* (death as perceived), and *le discours sur la mort* (the discourse on death).[3]

Linked as it is to the physical and biological phenomenon, *la mort subie* refers to the brute and material nature of death, its statistics, and its aetiology. This approach focuses on mortality rates and on the causes of death in a specific

[2] I am referring to John Searle's distinction (1995) between brute facts and social or institutional facts. Brute facts such as mountains, rivers, and weapons are not socially constructed and exist independently of human institutions. Instead, social facts are tacit agreements on the meaning of brute reality and need human practices in order to exist and to produce their effects. From this viewpoint, death is both a brute and an institutional fact.

[3] Vovelle's book *La Mort et l'Occident* has not been translated into English.

period and place and is concerned with establishing relationships between death and numerous variables such as wealth, age, sex, diet, job, habits, residential location, and the like. This is the death that can be measured through collection and analysis of quantitative data. Not only bio-medical and demographic studies fall into this category, but also the range of historical (Drévillon 2013) and statistical works (Lacina and Gleditsch 2005; Lacina et al. 2006) on casualties and lethality in war that can be found in International Relations (IR) scholarship. Conversely, *la mort vécue* is concerned with the rituals surrounding death and with the practices that accompany the sick through agony to the grave and even beyond burial. As a result, it refers to those disposal and funerary practices as well as private and public rites of grief and bereavement that are more or less consciously designed to manage death. Such practices can provide useful information on how people understand death and their own mortality in different ages and across societies. Finally, *le discours sur la mort* refers to the intellectual arguments employed to justify and explain death, such as theological, philosophical, and scientific discussions and representations of death and the afterlife that can be found in religious speculation, novels, art, medical, and scientific studies. Alberto Tenenti's study (1989) on death in the writings of Erasmus, Bellarmine, Calvin, and in the Italian Renaissance is a classic, masterly illustration of this approach.

As this investigation is concerned more with death as it is experienced and understood in particular historical periods than with the physical and material aspects of mortality, the main focus will be on what Vovelle termed *la mort vécue* and *le discours sur la mort*, but with a preference for the former. For, though the social meaning ascribed to death is ideational in essence, it does generally materialize in rituals, symbols, and daily practices. Intellectual discourse does not necessarily mirror social behaviour and attitudes towards death. As John Bowker (1991: 42; see also Black 2010: 87) observed, 'the human exploration of death occurs much more in such enterprises as liturgy, ritual, music, art, architecture, monuments, poetry, than it does in books about death'. However, a study on the changing meaning of death in war needs to go beyond the social and cultural history of death and include the symbolic strategies that western countries deployed to exploit death as a political instrument—strategies, as we shall see, that will succeed in the short term, but that will eventually fail in the post-Second World War era.

The empirical analysis to be found in the following chapters benefits also from the theoretical contribution provided by the flourishing sociological and anthropological literature on *the body* and death. In this fast-growing area of study, many investigations focus on the psychological relationship between the living and the dead, on the dying process, on grief reactions, and so forth. Other studies look at more political issues, such as the role of death in the construction of social order (Verdery 1999; Crossland 2000), as a key element in the building process of the colonial state (Mbembe 2003; Lomnitz 2005), and as a tool to reinforce racial segregation (Dennie 2009; Black 2010: ch. 2, 'Nazi Ways of Death').[4] Attitudes towards death and the dead have also been construed to explain social and power structure (Morris 1992) and to account for charismatic personality in political and religious leaders (Aberbach 1996), besides being described as one of the central features in the ideology of the European political Right in the first half of the twentieth century (Jesi 1979).[5]

Within this multifaceted field of analysis there is also a growing body of work inspired by Michel Foucault's notion of biopolitics (2003), Giorgio Agamben's idea of thanatopolitics (1998), and Achille Mbembe's concept of necropolitics (2003), all of which show how one of the crucial factors behind modern sovereignty is the creation of various types of subjectivities characterized by different degrees of exposure to death (Doty 2011; Robben 2014, Ferrándiz and Robben 2015; Squire 2017). Although this important scholarship looks at how lives are framed as *disposable*, its perspective can also help us to understand what can be seen as the reverse process through which western soldiers have been turned from expendables into individuals whose lives must be protected in war.

The Two Bodies of Soldiers

As mentioned in the Introduction, the narrative in the ensuing chapters is theoretically underpinned by Ernest Kantorowicz's notion of the 'king's two bodies', which provides a convenient starting point to explore the multiple meanings of death embodied in soldiers' employment in war, corpse disposal and burial practices, and war memory.

In his study on medieval political theology, Kantorowicz (1997) famously argued that the concept of the king's two bodies, a natural one that suffers and dies and a social (or institutional) one that transcends life and death and operates as a symbol for the political function, ensured the continuity of the monarchy even after the physical death of the king. The expression 'The King is dead. Long live the King' captures the notion that monarchy persists because

[4] For a critical overview of new developments in the historical literature on death, see Malone (2019a).

[5] 'Beautiful death' and 'Long live death!' were, respectively, the meaningful mottos of Italian and Spanish fascists. Carlo Mazzantini's autobiographical novel *A cercar la bella morte* (*In Search of a Beautiful Death*) (1995) is an example of this attitude.

the social body of the king never dies. Despite the demise of the corporeal body of the monarch, political continuity was preserved by the continuity of his deathless institutional body. As the great English jurist Edward Coke put it, 'the mortal king was God-made, but the immortal King, man-made' (in Kantorowicz 1997: 423). In other words, as the social body is not part of nature, it has the potential of surviving the death of the physical body.

Eschewing judgement on Kantorowicz's political theology and the original goal behind the notion of the 'king's two bodies' (namely, explaining the continuity of political power in the face of the monarch's death),[6] I suggest that this concept can be employed as a useful analytical tool in order to study the changing attitudes towards soldier deaths. Like a king in medieval Europe, though with obvious differences, a soldier may be conceived as having both a physical and an institutional body. The former refers to the corporeal properties of soldiers as living and functioning combatants. The latter includes the social aspects of soldiers' bodies after their death. Because the precise nature of the relationship between the social and natural body of soldiers remains largely undocumented, each historical section will offer a reconstruction of these two dimensions based on the three types of evidence mentioned above.

Soldiers' employment in war sums up how the *natural bodies* of soldiers have been handled throughout history. How troops are used on the battlefield can be meaningful for an assessment of the importance that states and societies attach to the life of their soldiers. Battlefield tactics, in particular, provide important hints on how states value the life and death of their combatants. Thus, the way soldiers are used in war can shed light on their degree of expendability and, in turn, on how their life is regarded and understood. Accordingly, the battlefield is the place where the research begins in the early modern period (Chapter 2, 'Bare Death') and where it ends in the current epoch (Chapter 4, 'Irrecoverable Death').[7]

The study of disposal and funerary practices, on the one hand, and commemorative devices and functions, on the other, proves to be useful to reconstruct the *social body* of soldiers—namely, the particular image attaching to soldiers after their natural bodies have ceased to function. It seems almost self-evident to point out that the ways soldiers' corpses are disposed of and their memory recollected can be quite revealing about how soldier deaths are valued and about the role soldiers were assigned in society.

[6] For a refined assessment of Kantorowicz's study, see Whalen (2020).

[7] That is why the empirical narrative in Chapter 2 develops from battlefield tactics to body disposal and war memory, while in Chapter 4 it starts with memory and body disposal and ends with how soldiers are used and not used on the battlefield.

At this point one may reasonably ask why the author has decided to focus on these three kinds of evidence. In fact, in order to reconstruct attitudes towards the dead, one could consider a wide range of alternative sources, such as obituaries (Fowler 2005; Fowler and Bielsa 2007; Phillips 2007), letters of condolence (Faust 2008), last will and diaries (Vovelle 1983; Brown 2008), medical journals and autobiographies (Strange 2005), theological writings (Chaunu 1978), art, iconography, and iconology (Huizinga 1922: ch. 11, 'The Vision of Death'; Panofsky 1964; Tenenti 1989), epitaphs, or a combination of them (Ariès 1981; Faust 2008; Laqueur 2015). This very diverse body of evidence could be the subject of this book, but such a diversity also points to an inherent danger that the researcher faces when confronting a potentially limitless documentation: focus on too wide a range of sources for studying long-term trends would inevitably end in an anecdotal approach. Thus, narrowing the range of sources was a necessary step in organizing a workable empirical canvas of research. Most importantly, the choice of concentrating on the three above-mentioned empirical sources derives from a simple fact: not only does a focus on soldier deaths make studying the corpse and mortuary practices almost natural, but it also justifies concentrating on war memorials. Indeed, most of these commemorative devices are or are formed of statues, which are, in the words of anthropologist Katherine Verdery (1999: 5), 'dead people cast in bronze or carved in stone'. Thus, in portraying the dead, statues are a replacement for the natural body preventing the social body from vanishing.

Combat: Fighting, Killing, and Dying in War

Throughout history, methods of using soldiers on the battlefield and military tactics have been devised and chosen for a variety of reasons, such as the demographic structure, geography, wealth, political organization, equipment, military doctrine, and, notably, developments in technology and logistics. The use of armed forces and tactics, however, has seldom been the mere product of rational, cold cost–benefit reckoning. Not occurring in a social vacuum, war has always been more than a matter of military calculation (Keegan 2004: 3–12; Farrell 2005; Creveld 2008).

None recognized the intimate relation between cultural/social elements and warfare more than Carl von Clausewitz. When he famously stated that war is 'a continuation of political intercourse, carried on with other means' (Clausewitz 1984: 87), he meant not only that war should be driven by political objectives, but that politics in the broad sense influences many aspects of warfare

(Echevarria 2007: 4).[8] In particular, the Prussian general maintained that there is a close connection between politics and war, in the sense that war is often shaped by the politics and society in which it takes place: 'The aims a belligerent adopts, and the resources he employs, must be governed by the particular characteristics of his own position; but they will also conform to the spirit of the age and to its general conditions,' wrote Clausewitz (1984: 594). Although 'these things do not belong to war itself', they nevertheless influence how war is waged. According to Clausewitz, indeed, 'every age had its own kind of war, its own limiting conditions, and its own peculiar preconceptions' (Clausewitz 1984: 594), which emerge from the social and political environment. For example, when he described war as a 'remarkable trinity,' formed of political objectives, military instruments, and popular and social forces, he claimed that it was this third dimension that turned eighteenth-century warfare from limited to absolute and 'made the wars of the French Revolution so different in kind from those of Frederick the Great' (Howard 1979: 977).

The social nature of warfare was recognized by Samuel P. Huntington, too, who argued in *The Soldier and the State* (1957: 2) that the 'military institutions of any society are shaped by two forces: a functional imperative stemming from the threats to the society's security and a societal imperative arising from social forces, ideologies, and institutions dominant within society.' Both forces are crucial in determining the nature and features of a state's military arm and its way of war, but the second force has nothing to do with the configuration of international politics, existing security threats, and available material resources and capabilities. Actually, the 'societal imperative arising from social forces' may even be in open contradiction with the 'functional imperative stemming from the threat to the society's security.'

Over the last few decades an expanding scholarship has shown that military tactics and strategy are far from being the outcome of a rational and detached decision-making process, but are rather a product of historical and cultural variables (Legro 1995; Kier 1997; Thomas 2001; Farrell 2005; Lebow 2008; Malešević 2010a; Lee 2011). Even strategic calculations based on available resources, weaponry of the day, and knowledge of the situation are conditioned by several ideational variables, which are often tacit and thus taken for granted. The nature of technology, the amount of financial resources, and the size of the army all set the standards of war (Lynn 2011: 90), but combatants and

[8] In particular, Clausewitz traced the revolutionary transformation of warfare during the Napoleonic era back to changes in the political conditions in which France found itself at the end of the eighteenth century. This is why, when John Keegan (1993: 11–12) criticized Clausewitz for having left culture out of his philosophy of war, he largely missed the point.

weaponry can be employed differently by different cultures. As Colin Gray (2005: 120) put it, 'there is a trialogue among what technology permits, what politics requires, and what society allows'. A few examples from this scholarship may illustrate how beliefs, norms, and ideas (that is, culture in its anthropological meaning) have impacted differently on the use of soldiers in war.

In his well-known study on the origins of the western way of war, military historian Victor David Hanson (1989) argued that limited warfare in ancient Greece before the fifth century BC, largely organized as brutal single pitched battles fielding spearmen organized in the serried ranks of the phalanx, was not the simple outcome of the human and military means available at the time, but was mainly a product of agrarian life and customs. Indeed, restraining organized violence in time and space was one of the main features of such warfare, the goal being to spare property and the lives of non-combatants. The war of the knights in the Middle Ages, with its chivalric code based on the concept of fairness and honour, which had the effect of limiting the degree of violence among noblemen at arms, is another remarkable example of how non-military variables might shape warfare in general and the use of force in particular (Rogers 2007; Cardini 2013).[9] In regard to warfare in the twentieth century, Stephen Van Evera's study (1984) on the 'cult of the offensive' before the First World War and Elisabeth Kier's analysis (1997) of French and British military doctrine in the inter-war period illustrate how military doctrine and warfare are strongly affected by a number of ideational variables such as culture, norms, and beliefs.

An important part of this scholarship has also tried to account for atrocities and barbarism in war towards both enemy soldiers and civilians. Ethnic identity (Kaplan 2005), religious fanaticism (Juergensmeyer 2020), ideological antagonism (Bartov 1992), military culture (Hull 2005), and other intangibles have been employed to explain why some wars or some fronts within the same war have been more brutal than others.[10]

These diverse and multifaceted works share a common background assumption: the social and cultural context is crucial to understanding the use of force, accounting for different military traditions, strategic preferences, and ways of war adopted by particular people and states. Moreover, this scholarship rejects

[9] As shown further on, these codes of military honour in relation to the enemy will persist in the early modern period, at least among generals and officers.

[10] There is also a major area of sociological and psychological literature that aims to explain both why soldiers fight and risk their lives in combat and why they break down (Marshall 1947; Shils and Janowitz 1948; Stouffer et al. 1949; Wessely 2006).

the view that the way combat is organized is determined merely by military factors. Indeed, such studies either focus on the social construction of killing in war or share the idea that some of the most important transformations in military practice have been influenced by changes and factors that are not military in nature.

Like many of these works, the present research starts from a broad historical and social view of warfare, which regards military methods and mores as something neither written in stone nor solely determined by material reality. In other words, ways of war vary according to the changing nature of the societies that develop them; and, in order to understand them, one needs to focus on the social and cultural constraints and enablers that underpin warfare. Quite surprisingly, however, this expanding body of manifold scholarly studies on the relationship between ideational variables and warfare seems to take no interest in why western countries have stopped regarding their soldiers as fighting tools that can be easily expended. In the following chapters a description of how combatants are employed in each period is given. The idea underlying the analysis is that the use of soldiers on the battlefield is revealing about how the life of the individual soldier is valued. Yet, it would be misleading to infer that this will turn out to be a comprehensive history of military tactics. Such a task is not only beyond the scope of this research but would be largely redundant. What is being attempted here is to show the varying degrees of soldiers' expendability resulting from the different types of force employment in the historical stages under scrutiny.

Besides the actual employment of troops on the battlefield, a remarkable factor that makes it easier to reconstruct the changing meaning of soldier deaths is the treatment received by the wounded in war. In a way, the wounded, especially the severely wounded, form an intermediate category of combatants between the living and the dead. Largely uncared for on the battlefield in the early modern period—medical treatment during combat was generally forbidden during military engagements (Gabriel 2013: 51), injured soldiers slowly but gradually became an object of both legal and practical protection. As we shall see, the second half of the nineteenth century was a period of great changes in the treatment of this human category.

Disposal of the Dead

While death is natural and inescapable, what societies do with the corpse is not. The body, one way or another, always goes back to nature, but how the transition takes place is a human decision usually rooted in culture. Whether the corpse is something important and meaningful or totally insignificant for a particular community is generally a social outcome, not a natural occurrence. The various rituals surrounding death, such as disposal and funerary practices, are eminently human actions rather than natural processes. Among these rituals, the disposal of human bodies before they decay is not only a basic need, but also a social and cultural turning point. With the disposal of bodies, the focus is shifted from what people were in society to what they will be for the community from which they have departed.

Although the biological reality of death is one and the same, a wide range of practices concerning human disposal can be traced across geographies and histories (Favole 2003). Even a cursory view of death customs across cultures or even within one society over a long period of time clearly shows the unlimited range of ways in which the disposal of bodies is organized. If one were to amass a detailed account of how corpses were treated (for example, abandoned, burned, roasted, toasted, eaten, buried, double buried, hung on trees) and preserved (mummified, embalmed, smoked, boiled, pickled, or dismembered), and how funerals were ritually organized in the past (as parties, sexual orgies, fights, or solemn, grave ceremonies),[11] one would probably refrain from drawing general conclusions.

Human societies have also been ambivalent towards the corpse, which has been seen in turn as a sacred body to be honoured and worshipped (Laderman 1996), as a commodity with monetary value from which to profit (Scheper-Hughes and Wacquant 2002), as polluting and burdensome trash to be discarded (Bloch and Parry 1982), or as a natural part of nature that should simply return to the earth and foster the food chain (Plumwood 2008). Furthermore, the various ways people have understood death (for example, as a mere rotting process, as a welcome transition, as a temporary sleep until resurrection, or as part of a process of reincarnation) are full of contradiction and inconsistency.

Although studying disposal practices might therefore be a difficult task, how corpses are dealt with can provide some evidence as to how the living relate to their dead. While arranging a proper burial is not necessary for the dead themselves, disposal and funerary practices imply that it is at least important to the living. As scholarship on the corpse has recently shown, the study of corpses is a potentially rich avenue for social and political enquiry. Especially today, as J. L. Foltyn (2008: 99) pointed out, the 'human corpse, and its social

[11] On some of these practices, see Ragon (1981) and Metcalf and Huntington (1991).

meanings and how it should be valued, discussed, disposed of, imaged, and used, is a critical subject, generating public debate, enormous media attention, and corporate interest'.

Beyond Death as the 'Great Leveller'

In order to understand the importance of disposal and funerary practices in reconstructing the social meaning of death, it is necessary to move beyond the idea of death as the 'Great Leveller'. Apart from emergency situations such as epidemics and other calamities when corpses pile up, death has never been a condition for social and political differences to be erased. While everyone is doomed to die and the socially undifferentiated universality of death cannot be denied—as the imagery of the *Danse macabre* reminded all human beings in late-medieval and early modern Europe[12]—mortality has always affected people unevenly. Moreover, and interestingly for the present research, the way the body has been disposed of and the related funeral rites have all historically been unequal. Undoubtedly, as the eminent literary critic Robert Pogue Harrison (2003: p. xi) claimed, to 'be human means above all to bury',[13] but how burials have been carried out and arranged has varied greatly throughout history and across societies. With few distinctions, burials and funerals have always reflected the social identities and walks of life of the deceased and have been employed to commemorate their worldly success or failure. As Stephanie Spars (2014: 103) argued, 'in the nature, arrangement, and spatial relationships of the material evidence at a burial site are representations of attitudes and behaviours of the living toward the dead, be it friend, enemy or unknown victim'.

The social esteem a person enjoyed is evident from the type of burial and the related last rites of passage. According to Ian Morris (1992), for instance, a precise relationship could be established in ancient Greece between type of burial and social standing. In early modern Europe as well, the position in the social hierarchy occupied by the departed could be inferred from the grave and the procession. Not only were disposal and funerary rites specially designed with due regard to social stratification, but they were orchestrated 'to display and reinforce the social distinctions of the dead' (Llewellyn 1991: 60). Funerary

[12] The *Danse macabre* (Dance of Death) and the image of the skull in general were employed as a reminder of equality in death in the face of a highly hierarchical society.

[13] In the *New Science* Giambattista Vico maintained that burying the dead is one of the essential features of human nature. In particular, he claimed that the connection between humanity and burying the dead is inscribed in language, since humanity 'had its origins in *humare*, to bury' (in Sherman 2014: 11). However, it should be noted that many populations, such as the Yafar from New Guinea, the Djour from Sudan, indigenous people from the Solomon Islands, and others, left their dead to animals and insects. On these practices see Ragon (1981).

rites were also intentionally employed to reinforce the social identity of a particular class, as in the case of the support system organized to manage funerary rituals by the trade unions in nineteenth-century Britain (Laqueur 1983: 117).

In a sort of horrific mirror image, mutilation and dismemberment of lifeless bodies were conceived as a form of ultimate punishment that went beyond killing of the victim (De Luna 2006). In Sophocles' tragedy *Antigone*, Creon's initial decision to leave the corpse of Polyneices 'unburied so men may see him ripped for food by dogs and vultures' (Sophocles 1989: 28) was a political judgement and a posthumous conviction on someone who had betrayed his fatherland (namely, Thebes). The posthumous execution and treatment reserved by English monarchists for Oliver Cromwell's disinterred corpse was a further well-known macabre example of the genre (McMains 2000: 20). After Cromwell had been disinterred from Westminster Abbey, his corpse was hanged and beheaded. Then, the head was publicly shown and the body thrown into a pit. The actions of disinterring Cromwell's corpse, spreading it out in full view, and preventing its reburial were all meant as a sort of post-mortem punishment.[14]

By the same token, the denial of funeral rites, as for Nazi war criminals after their execution and cremation in the aftermath of the Nuremberg Trials, was employed both to obviate a place of remembrance for future Nazi sympathizers and to condemn the dead person's behaviour in life.[15] In other words, as these examples illustrate, the disposal of bodies has often meant a final judgement on the deeds of the deceased.[16]

Hence, to understand the meanings of death rituals, we must reject the idea of death as the Great Leveller: disposal practices reflect the implicit but visible status of the individual at different times and across societies. Likewise, as we shall see in the following chapters, they appear to provide evidence for the 'changing status' of the soldier in the western world.[17] Indeed, it might be suggested that the way the corpses of the war dead are dealt with is a reflection of their social positions. As shown further on in the book, the way soldiers' bodies

[14] Another (in)famous example of this genre is that of the French revolutionary Mirabeau. From being the 'inaugural corpse' (Legacey 2019: 26) of the Panthéon, his tomb was desecrated and his cadaver thrown into the sewers of Paris when after his death it was discovered that he had held secret relations with Louis XVI.

[15] The refusal by local religious leaders to give a Muslim burial to Adel Kermiche, the Jihadi attacker who killed a Catholic priest in the French city of Rouen (26 July 2016), is a recent example of post-mortem punishment.

[16] Until a few years ago, in the graves of executed prisoners located in the cemetery of Huntsville, the site of Texas's Execution Facilities, names were not carved on the grave markers. Only the date of death and their Department of Correction (DOC) numbers were listed.

[17] On the notion of the 'changing status of the corpse', see Laderman (1996).

have been disposed of has radically changed since the sixteenth century. From being considered not unlike animal carcasses that needed to be discarded only for hygienic reasons by burying or destroying them, soldiers' corpses are now treated with great care and have become the object of clinical and biomedical screening, such as fingerprinting, dental examination, DNA tests, and other sophisticated forensic methods.

Cemeteries as Cultural Texts

Alongside disposal practices, important clues are given by cemeteries and grave markers as well, which, despite differences among cultures and societies, remain loci for mourning and remembrance. Cemeteries are complex human and cultural creations, whose location, shape, inscriptions, and symbols express certain attitudes towards the dead.

Since not every place of disposal is a cemetery and interment might take place without funeral and ritual practices, the distinction between burial and cemetery should be stressed and clarified. On this point, Jacek Kolbuszewski distinguishes the burial place as the mere site where cadavers are interred from the cemetery as a specific space created with the additional purpose of hosting funeral practices. In particular, Kolbuszewski defines the cemetery as 'a certain sector of space delimited by certain a priori formulated resolutions, according to which it is there that funeral practices consistent with religious, ethnic, cultural (that is customary) and other easily defined needs of a given community, will be carried out' (in Rugg 2000: 260). In other words, a cemetery not only serves for interment, but also provides survivors with a spatial locus on which to focus emotions and grieving practices. This is why, as Eva Reimers (1999: 150) clarified, a cemetery can be read 'as a cultural text about society and the individuals that have found their last place of rest in its burial lots'.

As we shall see at the end of Chapter 2 ('Intermezzo'), putting the dead to rest in individualized burials in a cemetery is a modern spatial invention dating from the late eighteenth and early nineteenth centuries (Laqueur 2001: 3). Before this period, the location of graves was among the living. Cadavers were buried mainly in or near churches, which were located in turn at the centre of communities rather than outside the city walls. Moreover, churchyards were even used for gambling, carrying on business, public recreation in general, and even for the grazing of animals (Tarlow 1999: 111–14), practices that would be unimaginable and disrespectful today, and that slowly and gradually began to be outlawed. The innovation of the modern cemetery simultaneously protected and segregated the new community of the dead from the world of the living and was one of the factors that led to a new relationship between the dead and the survivors. For, with the introduction of modern cemeteries, family bonds could be manifested despite death and individual loss.

Kolbuszewski's above-mentioned distinction between a mere burial place and a cemetery is especially relevant for burial practices regarding soldiers. As we shall see, in early modern Europe soldiers were generally interred in mass graves excavated in the proximity of the battlefield. It was not until the second half of the nineteenth century that the practice of burying ordinary soldiers in individual graves in military cemeteries with a record of their names emerged and gradually developed.

'Who wants flowers when you are dead?': The Social Functions of Funerals

Among death ritual,[18] funerals are complex events that might range from a purely private rite of mourning, involving family and friends, to a public ceremony meant to affirm particular values such as authority, political hierarchies, and social bonding.[19] Although the dead person is at the centre of the funeral, the focus is directed from the deceased to the world of the living. As an early critic of funerary practices put it: 'The pomp of the burial, with great accompaniment, and the sumptuous funeral are more for the consolation of the living than for the favor and aid of the dead' (in Lomnitz 2005: 101–2). Especially since the modern era, funerals have been organized more to reassure the bereaved rather than to stand by the dead. In particular, funerary rites have been used both to facilitate acceptance of the end of the natural body and to establish and sustain a particular memory of the social body of the deceased.

That is especially true with state funerals, which are theatrical rituals with important political effects. They are generally characterized by solemn ceremonies exalting heroic and civic virtue and the deeds of great men or women. By bringing the nation together, these *fêtes funèbres* are intended to create unity, to renew the social pact, and—as an educational tool—to inculcate values of national identity in the citizenry. As French statesman Léon Gambetta famously declared, funerals are 'the most propitious occasion for the education of the people'. In this regard, the nationalist writer Maurice Barrès grasped the full potential of the dead when he talked about the 'social virtues of a cadaver' (in Varley 2008: 73). Indeed, from Pericles' oration down to the present, the

[18] The quotation in the heading is from Holden Caulfield, in Salinger's *The Catcher in the Rye* (2014: 172).

[19] The importance of funerals is also borne out by the fact that they provide room for bitter political contests and resistance and have been exploited as occasions to express political hostility. One of the most famous examples in French history was poet 'Paul Déroulède's attempt to turn troops and the crowd' against the French government during the 'rites for President Faure' on 23 February 1899 (Rearick 1986: 406).

pedagogy of funerals has always been an effective means of reinvigorating the spirit of a polity (Laderman 1997).

In relation to the present research, it should be noted that, whereas state funerals in the early modern period centred on well-known 'great men', in the second half of the nineteenth century they were extended to ordinary soldiers. As we shall see, since the second half of the nineteenth century, soldiers' funerals have been conducted according to an ambiguous combination of civilian and military rites, the former originating in funerary practices of society at large and the latter in pompous patriotic pride based on nationalist ideology. Even on the battlefront of the First World War, civilian funerary rites were creatively adapted to the emergency situation created by war with the goal of respecting the mourning practices adopted by the population in peacetime. As Chapter 4 will illustrate, states accepted the societal demands for peacetime civilian practices concerning the dead and transferred civilian disposal customs and funerary rituals to military casualties.

The disposal of the corpse, interment in cemeteries, and the rites and rituals that accompany the dead are only a part of the ceremonial practices that lend special meaning to the dead. Memory and remembrance are the other side of the coin.

Memory and Remembrance

If death implies, among other disturbing things, the risk of being forgotten, then memory and remembrance are its imperfect remedies. While death is irreversible, the departed can be brought to new life in the domain of recollection. For, one of the goals of storing something in the memory is to keep it alive against the eroding effects of time. As Susan Sontag (2003: 115) put it, memory 'is achingly, the only relation we can have with the dead'.

Here we are principally concerned not with private remembrance of personal loss, but rather with how society collectively interprets death—that is, with communal loss. The main object of investigation is not, therefore, how soldiers made sense of the encounter with battlefield death,[20] but rather the collective and societal representation of military casualties. Accordingly, we are interested in what Maurice Halbwachs termed 'collective memory'. The French sociologist distinguished between individual, private memory, on the one hand, and collective memory, on the other, emphasizing that collective

[20] On how soldiers understood battlefield experience, see Harari (2008).

memory cannot be reduced to the sum of individual memories (Halbwachs 1992: ch. 4), 'The Localization of Memories'). According to Halbwachs, individuals remember in the literal and physical sense, while groups construct collective memory by determining what is and is not memorable through a process of remembrance and forgetting. What is excluded from the memory of a particular community is rarely due to mere chance (Lowenthal 1985: 204–6; Wood 2000: 145); more often, disregarding certain details of the past is the result of either specific political interests or widespread social exigencies. By including and excluding certain events, subjects, and actions, collective memory does not simply recall past events, but has the effect of shaping the present (Lowenthal 1985: 27).[21] Because collective memory mirrors both the social context of a particular community and the power struggle behind recollection of the past, the present study is also geared to understanding how these two different factors have contributed to the meaning attached to the war dead from the early modern period to the present era.

This part of the analysis will benefit from the fact that war memory has become the subject of considerable scholarly interest and extensive research in cultural history, sociology, and political science, as commemorative practices have proved to be a fruitful area of enquiry for the study of politics and society. In particular, the present study relies on a rich literature on memorials that began to emerge in the 1980s and to which even the most recent writings return. However, although the value of these studies is immeasurable, by focusing either on a single national case (e.g., Ignatieff 1984; Hynes 1990; Mosse 1990; Merridale 1999; Inglis 2008; Kattago 2009) or on a single historical period, often the First World War (e.g., Wilkinson 1997; Lloyd 1998; Williams 2009), they fail to place the analysis in a larger historical context. The main limit of these enquiries is their lack of a long-term perspective, resulting in a fragmented picture of the changing nature of war memory, which cannot trace and account for historical developments and ruptures. In order to avoid some of the problems connected with covering merely a brief period of time or a single country, and to understand shifts over time, the scope of this study will be broadened to include several countries over a long span. Such an

[21] Although Halbwachs's notion of collective memory is central to the present study as it clarifies the roots of the politics of memory, his 'ultimate opposition between memory and history' (Halbwachs 1980: 78), between imaginative reconstruction and a verifiable and objective science of history, is questionable not only because objectivity, even when an explicit goal of the historian, is hard to achieve, but also because history does not exist as something out there waiting to be discovered. Both memory and history are cultural products influenced by a variety of factors such as identities, political and economic interests, power, and so forth.

approach, albeit limited in depth and preliminary in its findings, may open up new avenues for the study of memory and its influence on politics and war.

War Memorials and Commemoration

Collective memory and memorialization include oral testimony, remembrance days, annual gatherings, public ceremonies, and the like. Memory can also be cast in statues, monuments, and memorials. Wars, especially, have been commemorated by a variety of material objects such as obelisks, arches, columns, statues, edifices, tombs, plaques, chapels, museums, bodily adornments, or, more recently, by utilitarian memorials such as schools, hospitals, and other public places built for human activities and for the improvement of social conditions. These artefacts and buildings are still among the most visible as well as spectacular constructions that remind us of a particular conflict. Being placed in public squares, crossroads, cemeteries, and churchyards, they are a central part of the urban landscape and space of many cities.

Among the many reasons behind the construction of such material objects, it is worthwhile quoting the eminent French philosopher Gaston Bachelard, who maintained that 'memories are motionless and the more securely they are fixed in space, the sounder they are' (in Bachleitner 2021: 1). And, paraphrasing Cambridge Egyptologist Barry Kemp (2006: 248), one may add that memory needs architecture for its fullest expression. Indeed, thanks to monuments and memorials, recollection is more than words or ideas; it becomes an object or a place that can be seen, touched, and visited.[22]

In terms of shape, it must be emphasized that war memorials and monuments are not built primarily for the sake of beauty, but rather in order to construct memory and meaning. In particular, according to students of commemoration, such memorials should be considered not as mere pieces of popular art (Gusfield and Michalowicz 1984; Brandt 1994), but as manifestations of a specific view of war. By marking special events with a particular shape, monuments and memorials express the values, ideals, and aspirations of particular people, states, or political regimes. Their aesthetics is certainly a product both of the artist's personal taste and creativity and of the prevalent styles of the epoch, but memorials are erected to communicate a specific interpretation of the war—something that no artist or architect can ignore. Thus, war memorials are material devices with the explicit aim of providing a well-defined reading of history and a way of making the past meaningful for the

[22] This is what Pierre Nora had in mind when he chose the phrase *lieux de mémoire* (sites of memory) as the title of the monumental research on French collective memory published under his direction between 1984 and 1992.

present. The iconography, the design, the inscriptions, and the visual representation employed are intended to signify a specific set of values and to create certain political and social messages. Thus, what we often view today as meaningless or as an object of some curiosity (Isnenghi 2005: 321) is, in the words of Sally Morgan (1998: 103), an 'apparatus of social memory'. Changes in the content and form of memorialization and in the commemorative landscape reveal that modifications are taking place in the social or political realms.

Most war memorials were built to glorify a victorious armed conflict; some of them, however, were erected to solidify a nation and, more rarely, to celebrate a people despite military defeat, as testified, for instance, by Serbian commemorative practices dedicated to defeat in the battle of Kosovo Polje in 1389 (Mock 2012: 157–61). Although war memorials have a variety of symbolic functions, one of the most important is to commemorate the war dead. As American sociologist Bernard Barber (1949: 65) maintained in a pioneering article published in *Social Forces*, war memorials reflect 'the attitudes and values of a community toward those persons and deeds that are memorialized'. Katherine Verdery (1999: 5) goes even further in declaring that, 'by arresting the process of that person's bodily decay, a statue alters the temporality associated with the person, bringing him into the realm of the timeless of the sacred, like an icon'. In other words, statues conceived as monumental bodies attempt to perpetuate the memory of their subjects and provide the dead with a specific meaning.

There are two major questions, though. Which of the dead should be revived and why should they be remembered? A key distinction is made between the dead who are to be taken account of and named and those who can be left anonymous and forgotten. Or, to borrow Judith Butler's wording (2004: 20; 2009), which lives are *grievable* in war? Who is commemorated and how commemoration takes place may reveal how the war dead are perceived in a particular society and the position soldiers occupied in the community. That is why war memorials should be regarded as material devices that convey not only different ways of viewing war but also different ways of understanding the value ascribed to the fallen. In particular, as Anthony King (2010: 3) clarified, 'the commemoration of the war dead … is charged manifestation of existing understanding and practice'. As a result, an enquiry into commemorative forms can offer important insights into how life and death in war are valued and understood.[23]

[23] As there is some overlapping between death rites and memorialization, it should be noted that in some cases—such as the tomb of the Unknown Soldier—the distinction between the two, especially between a memorial and a grave, is not always clear and might not apply at all. Not only may

In this connection, it should be noted that military deeds and war heroes seem always to have been celebrated and perpetuated in memory, whereas not all combatants in war were commemorated. Significantly for the subject matter at issue, the practice of remembering and honouring ordinary soldiers is not a typical western memorial motif, but a relatively recent phenomenon. While the main traditional commemorative theme used to be triumph in war centring on military and political leaders, in the second half of the nineteenth century the ordinary soldier entered the vocabulary of memorialization, more or less concurrently with the construction of military cemeteries. Interestingly, both victory and defeat would be commemorated as a representation of the war dead.

A second set of important questions has to do with the visual representation of the fallen: is death represented as glorious sacrifice or, on the contrary, as pain and unbearable loss? Is the soldier portrayed as a killing machine or as a victim?[24] Suffice it to say that the iconography of war has radically shifted from a representation emphasizing victory, aggressiveness, and the positive function of blood, to one focusing on the suffering of individual soldiers and civilians alike, which is a major break that will be described in Chapter 4 as testimony of the profound change in the meaning attached to a soldier's life in the present era.

Honouring the War Dead: A State Project or a Spontaneous Effect?

With reference to the war dead, one essential question underlying this study as a whole concerns the origins of military cemeteries and the commemoration of ordinary soldiers. These institutions and practices have been interpreted either as a state project aiming to foster nationalist sentiments in the age of the masses (Hobsbawm 1983; Mayo 1988; Mosse 1990) or as a mechanism of mourning and consolation meant to heal the wounds of war, especially for the families of the fallen (Winter 1995; Tarlow 1999).

Claiming that war cemeteries and commemoration of ordinary soldiers developed *from above* as an ideological instrument would equate to arguing that it was largely a state-controlled process manipulated by political leaders, nationalist intellectuals, and the media. Conversely, holding that it was an outcome *from below* that emerged from the need to mourn the fallen would amount to alleging that it was a spontaneous, natural effect of the mass killings

grave markers be considered memorials, but fragments of bodies like bones (in ossuaries) should be considered ways to memorialize the fallen.

[24] The French writer and officer Alfred de Vigny (1797–1863) rightly observed that the soldier is both a victim and an executioner (in Chamayou 2015: 103).

caused by modern warfare. In regard to this matter, Jay Winter and Emmanuel Sivan (1999: 9) make a distinction between collective memory and collective remembrance. The former is by and large a creation of the political and social elites; the latter, by contrast, relies on the agency of several groups with different memories in the process of recollecting and constructing the past.

Most military cemeteries and memorials were and still are state sponsored and have been exploited by governments to try and deliberately shape the collective meaning and memory of war. According to Gusfield and Michalowicz (1984), memorials are tools employed by states for creating and maintaining popular allegiance, and for mobilizing consensus and resources. In this regard, John Gillis (1994) has rightly noted that most frequently the individuals who decide what to commemorate and how to do so are those in power.

Unquestionably, memorialization has often been devised by states according to their changing political needs, but such an 'instrumental' perspective appears to be too narrow. Social reality in general and memory in particular are more complex than that. States are far from being able to manufacture and control the production of meaning in an absolute way.[25] Meaning often remains a contested and fractured terrain of struggle among different groups: its construction does not take place in a social vacuum but, rather, is shaped by what happens in society and reflects the prevailing values of a community. Cultural and symbolic representations rarely succeed if they are not already powerful in their own right (Taylor 1989: 151). In this light, official acts of commemoration are not only the expression of a political strategy from *above* but are also—largely, though not solely—the result of pressure from *below*, as noted by Winter and Sivan (1999). Moreover, the polarized reading of war commemoration as either a state-sponsored project or a natural act of societal mourning is based on two erroneous dichotomies: state versus civil society and political manipulation versus grief. In an attempt to prioritize two of these elements, scholars have reduced the possibility of understanding the interaction between both pairs. Actually—as I clarify in the following section, where I sketch an explanation of attitudes towards soldier deaths—there is no real contradiction between the reading of war commemoration as a state-centred process by eminent scholars such as Eric Hobsbawm (1983) and Benedict Anderson (1991) and the reading of commemoration as an expression of grief organized by 'individuals and social groups' suggested by Winter and Sivan (1999: 9). As even

[25] There are, of course, examples of war memorials that are completely controlled in their design and production by states. The Korean War Veterans Memorial, which I will discuss in Chapter 4 in relation to the Vietnam Veterans Memorial, is a case in point.

Hobsbawm (1983: 307), a supporter of the state-centred approach, acknowledged at the end of his study on 'invented traditions', symbolic policies work only because they 'meet a felt—not necessarily a clearly understood—need among particular bodies of people'. In other words, for the state to mobilize masses through commemorative practices, one needs to touch the pre-existing collective understanding of the people. While memorials and public mourning rituals are generally organized and controlled by states, this study will show further on that they often originated in society, were publicly sustained, and even financially supported by private individuals and societal groups. As we shall see, not only did the state contribute to creating monuments and memorials for the fallen soldiers, but societal groups (for example, veterans' and ladies' associations, families, churches) also helped to erect these forms of material culture. The implication is not that memorialization may be a simple product of social spontaneity, but that this kind of project is not completely controlled and owned by states and by those in power.

Explaining Changes in Attitudes towards Soldier Deaths

The birth of the military cemetery, the spread of individualized burials with the names of the dead marked, novel commemorative practices for ordinary soldiers, and current casualty aversion demand some kind of unified explanation.

At this point it should be clear that I do not think that democracy is the driving variable that explains changing attitudes towards military deaths. Neither the new social body of soldiers (namely, individualized burials and the incorporation of ordinary soldiers in war memory) nor casualty aversion is historically linked to the democratic regime. Similar disposal practices developed both in a democracy like the United States in the aftermath of the Civil War and in an autocracy like Prussia in the Franco-Prussian conflict, respectively. Although this comparison alone is far from sufficient to show that the political regime is not at the origin of current mortuary practices concerning ordinary soldiers, the fact that the so-called democratization of war remembrance would later be embraced by a dictatorship like Fascist Italy casts a long shadow of doubt on the impact of democracy. The modern cult of the war dead ran not only cross-country but also cross-regime. Moreover, casualty aversion, too, is historically unrelated to democracy. True, at first glance the explanation emphasizing democracy's aversion to the prospect of casualties seems to account for the increasing sensitivity to the war dead. By promoting moral and political equality among citizens, regardless of social origin and identity, democracies appear naturally averse to the death of their citizen-soldiers. However, this explanation does not stand up to historical scrutiny for a simple chronological fact: while casualty sensitivity is largely a post-Second World War phenomenon, democracy is not. The cases of Britain and France in the First World War are striking examples of democracies that during the conflict were by no means averse to their soldier toll—a fact that clashes with the commonsensical idea that the political regime *in itself* influences attitudes towards military casualties.[26] Although the process of individualization and democracy are historically, socially, and politically connected, they are not the same phenomena.

One further common explanation employed to account for casualty aversion refers to the so-called *demographic transition*. According to Edward Luttwak (1994), the changing demographic composition of western societies is the key variable explaining casualty sensitivity. Since families have fewer children now than in the past (a decline in fertility rate) and children have high expectations of reaching adulthood (Colombo 2021: 17), human life in war is considered more valuable today than in previous epochs. This, runs the argument, is why western states and especially policymakers have become reluctant to risk the lives of their young in war. However, taking the United States as a test case, it should be noted that the real decrease in the number of children per family occurred long before friendly casualties began to be a sensitive issue in the post-Second World War period. Although it is true that there has been a considerable overall fertility decline over the last two centuries—from about 7 children per American woman in 1800 to 1.705 in 2019—the most significant decrease in children per American woman occurred when military casualties were not an issue at all. Thus, while the average American woman of childbearing age had about 5.3 children in 1860, in 1915 the number decreased to 3.5 and to 2.5 in 1940, with an interesting increase throughout the 1950s (3.5 in 1960). If demography were able to account for casualty aversion, then in the 1960s American society would have been less casualty sensitive than in the 1930s and 1940s and as casualty sensitive as in 1915. But, as we know, that is not the case. In relation to life expectancy, it is noteworthy that, although there has been a dramatic change in life expectancy at birth, such a change is significantly less dramatic for men at military age—for those individuals, in other

[26] In the course of the book, I present and critique other alternative explanations for the phenomena here considered. On casualty aversion as an outcome of the type of recruitment, and nationalism and the magnitude of the war dead in the First World War as a source of modern war memory, see Chapter 3.

words, who survived the ravages of disease and malnutrition, which for most of human history contributed to high child mortality rates. Keeping to the US context, whereas life expectancy for men was 38.3 at birth in 1850, 47.88 in 1900–2, 65.47 in 1949–51, and 66.80 in 1959–61, the figures at age 20 were 42.9 in 1850, 42.03 in 1900–2, 48.92 in 1949–51, and 49.77 in 1959–61 (Arias 2019), which means that life expectancy at military age increased by less than 7 years from 1850 to 1961.[27] Hence, demographic changes seem incapable of fully explaining the radical transformation in attitudes to soldiers' lives in the post-Second World War period.

Proponents of the so-called CNN effect and the Dover test in turn postulate that images of the fallen and body bags returning home are shocking to the public, who react negatively in terms of support for war. However, it should be emphasized that dead bodies have not always shocked the public. First, images of war dead and mangled bodies are not a post-Second World War innovation. The American Civil War (Rosenheim 2013), for example, produced large amounts of photographic material without causing any attempts to minimize casualties in subsequent wars. Although Matthew Brady's famous New York exhibition *The Dead of Antietam* (1862), which displayed the bodies of the fallen in the bloodiest battle fought on American soil, offered civilians images of the brutal reality of war and horrified visitors,[28] it did not make the North shy of expending men. Moreover, during some conflicts images of the war dead were used by governments to boost the home front—for example, when President Roosevelt allowed *Life* magazine to publish graphic images of killed US service members in September 1943.[29] Finally, the role of televised images appears to have been greatly overestimated. Doubtless, television and now the Internet bring images of war and death into living rooms with an unprecedented frequency, but we should also note that, 'whatever impact television had' during the Vietnam war, there were no great differences in public reactions to casualties between this conflict and the Korean War (J. Mueller 1973: 167), though the former was the first *television war* and the latter was reported mainly through radio and newsreels (K. P. Mueller (2000).

[27] For another sceptical view on Luttwak's demographic explanation, see Feaver and Miller (2014: 150–3), who suggest that changes neither in neonatal mortality nor in birth rates can account for the increasing casualty sensitivity in American society.

[28] The *New York Times* (1862) described the exhibition in the following way: 'Mr BRADY has done something to bring home to us the terrible reality and earnestness of war. If he has not brought bodies and laid them in our dooryards and along the streets, he has done something very like it.'

[29] On how these depictions of the war dead gained public acceptance during the Second World War, see Kimble (2016).

Finally, according to Gelpi (2006) and Gelpi et al. (2009: ch. 8,'Conclusion'), casualty aversion originates in the evolution of military technology. In particular, Gelpi (2006: 142) suggested that 'the advance of technology has changed the level of "necessary" casualties that the public believes success requires ... the technological prowess of the American military has changed the public's expectations of how many casualties are ... needed to achieve victory'. The explanation of casualty sensitivity as a result of advances in military technology suffers from a different but no less significant weakness. Undoubtedly, in some types of conflict, technology makes it possible to have zero or very low casualties. However, it is not clear why sensitivity to casualties kicked in during the Vietnam War and, especially, in Korea, when expectations of low casualties could hardly exist in the light of the high lethality of the respective previous conflicts. Advanced technology may explain both why a risk-averse warfare can be conducted and why expectations of low casualties increase, but it seems less capable of accounting for the origins of western societies' sensitivity to their war dead. Accordingly, technology is not the critical factor in this story. Although it allows us to do things that were not only impossible but utterly unthinkable a few decades ago—such as projecting military power without projecting human vulnerability—sensitivity to casualties began in a world with no combat drones or precision weapons.

A Unified Explanation of the Changing Meaning of Military Casualties

In order to account for the changing meaning of military deaths in the western world and mindful of the limits of current explanations, I would advance the following proposition. The rise and spread of the modern conception of the individual initially modified practices, gestures, and values attaching to the dead in general and then transformed attitudes towards soldier deaths. The impact of this macro-variable was not linear, however, and occurred at different stages. In particular, the impact of the process of individualization was mediated by a series of policies devised and implemented by the most important political actor of the modern era, the sovereign state, which in the short term successfully channelled the process of individualization into the holistic narrative of the nation.

In a highly schematic way, we may outline this causal process in the following terms. First, societal attitudes to death changed towards an individualization of the dead, visible both in the structure of the new cemeteries built in the late eighteenth century and in the new importance attached to the departed (see Chapter 2, 'Intermezzo'). Second, because western states could not disregard the novel relationship between the living and the dead, they brought mortuary practices concerning soldiers and war commemoration into line with the values of the time. From the mid-nineteenth century on (see Chapter 3, 'Sacrificial Death'), the change began to impact on the social body of soldiers with the construction of military cemeteries where soldiers could be laid to rest (when feasible) in an individualized manner and with the inclusion of ordinary soldiers and their individual identity in commemorative practices (for example, the great lists of names on memorials). However, the process of individualizing soldier deaths had limits. Although in the aftermath of wars states adjusted mortuary and commemorative practices to the novel sensitivity towards the war dead, they represented the fallen soldiers through a narrative based on national heroism that justified mass sacrifice on the battlefield. From this viewpoint, the cult of the fallen soldier can be considered as an attempt to manipulate the attitude towards death that developed over the nineteenth century. Thus, the modern memorialization of ordinary soldiers was both a spontaneous and a manipulative process emerging through the dynamic interaction between societal forces and public authorities. In this sense, the opposing interpretations of modern war commemoration reached, on the one hand, by Hobsbawm (1983), Mosse (1990), and Anderson (1991) and, on the other, by Winter (1995) and Winter and Sivan (1999) are consistent if they are applied to explain different aspects of this story. The will to remember the sacrifice of common soldiers originated in society, in particular; it resulted from pre-war attitudes towards death at the societal level; but how military cemeteries were organized and commemoration was shaped resulted both from popular demand (individualization of burials) and from a state desire to foster nationalistic messages (the rhetoric of sacrificial death in war). From the state's viewpoint, this was a strategy that worked in the short but failed in the long term, because the full emergence of the individual contradicted the holistic narrative of the nation. Paradoxically, states' incorporation of ordinary soldiers in commemorative practices contributed to the sacralization of soldiers' natural bodies. While commemoration was meant to foster the idea that combatants died for something that outlasted their individuality and thus naturalized them as sacrificial beings, the long-term effect was to validate the process of individualization, making it hard to justify their death. It is the irony of late-nineteenth-century and First World War commemorative practices that governments should have employed them to justify the war effort, but in so doing they conjured up a view of death in war that would turn against them in the following decades.[30] In the event, this new attitude would redefine western states' obligation towards their soldiers, as well as the nature of western warfare. As it happened, after the Second World War full development of the process of individualization of soldier death impacted not only on the social body of soldiers but on their natural body, with the rise of casualty aversion.

On the basis of the explanation provided here and the threefold bloc of evidence discussed throughout this chapter (namely, force employment, disposal of the dead, and war memory), we can begin to make sense of the way soldiers have been valued in the modern and contemporary ages. The broad trends that have shaped the meaning of soldier deaths in these periods can now be described with details of changes and continuities.

[30] From this viewpoint, Thomas Laqueur's intuition (2015: 31) that death practices do not simply reflect culture but that they create it captures the role of states' commemorative and mortuary practices in this story.

2

Bare Death

Soldiers as Mere Fighting Tools

In one of the Reith Lectures, military historian John Keegan (1999a: 43–9) remarked on the high standing veterans enjoy in Britain today. He noted that soldiers are not only respected as heroes; they are often portrayed as victims of war. Keegan then compared this attitude with past views of soldiers and concluded that a revolution in social perception had occurred: from being equated with the social status of 'prostitutes and criminals', soldiers are now regarded as honourable professionals deserving the highest admiration. Such a description not only appears to apply to British soldiers but can be extended to western soldiers in general. Indeed, while there is often disagreement on the merits of particular wars, most Americans and Europeans now agree that soldiers deserve respect for choosing a risky profession for the potential benefit of the entire nation.

This perception is worlds apart from the public view of the private soldier in the early modern period; then the soldier was seen as an unmitigated scoundrel, a drifter, a thief, or a social drop-out at best. This disparaging image was partly deserved. In early modern Europe, armies were formed of a motley humanity including experienced mercenaries, volunteers, a small portion of conscripted recruits, and an assortment of marginalized and disadvantaged people seeking to make a living by supplementing a steady salary with plunder (Woloch 1982: 53). Enlisting was often an alternative to prison, to an empty belly, a way to escape from a girlfriend's full belly, and from other less than honourable activities (Holmes 2005; 226; 2011: p. xxv). Although not all soldiers in early modern Europe came from the lowest social strata (Corvisier 1979), it is no overstatement to say that soldiering was considered an occupation for men who had little to offer to society.[1] As military camps were often

[1] On unemployment as one of the main reasons for enlisting in the British army under Wellington, see Coss (2010). Even today there is a negative relation between income and recruitment, especially in times of economic depression. This is what one may call 'economic conscription'. For example, current data still confirm the special vulnerability of lower-class soldiers in the United States. On American wartime inequalities, see Kriner and Shen (2010).

Life, Death, and the Western Way of War. Lorenzo Zambernardi, Oxford University Press.
 DOI: 10.1093/oso/9780192858245.003.0003

accompanied by a large number of prostitutes and a few professional thieves, a soldier's life, at peace or war, was likened to a criminal's. Not surprisingly, the military were vilified by the rest of society, and the low regard in which they were held often debarred them from public places, especially taverns.

Furthermore, soldiers' exploits on and off the battlefield did not help them win the respect of society. They looted, blackmailed, and robbed civilians if need be or, more probably, whenever they could. They also tortured and killed if they did not get what they demanded. In some instances, soldiers were housed in the homes of civilians as a form of punishment and intimidation for the latter, such as in the case of the *dragonnades* when in 1681 Louis XIV decreed that dragoons were to be billeted in Huguenot families.

Soldiers' behaviour was often a simple result of their social origins. The recruitment of privates was not restricted only to the non-productive elements of society (Parker 2004: 46–7). Occasionally, convicts and criminals were also drafted in order to increase numbers (Conway 1987), and they could be granted a pardon by enlisting. One of the reasons why so many soldiers used to wear long hair was to hide their ears, which had been cropped as a form of physical punishment for a variety of crimes (Hale 1985: 86). Even in late-eighteenth-century America, the Continental Army of the Revolution drew its forces from the poorest ranks of society and comprised a large number of convicts, drifters, and drafted substitutes (Kohn 1981: 557).

Not surprisingly, those who wrote about soldiers could not help depicting them and their lives in harsh terms. Louis de Jaucourt (1704–79), a major contributor to the *Encyclopédie*, recorded that 'soldiers in the countries of Europe are truly … the most vile portion of the subjects of the nation' (in Lynn 2011: 96). Edmund Burk described British soldiers as 'rapacious and licentious' (in Snape 2005: 13). Claude-Louis de Saint-Germain, a French general and later, under Louis XVI, Minister of War (1775–7), pointed out that 'the army must inevitably consist of the scum of the people' (in O'Connell 1989: 153). In similar words, the oft-quoted statement by Arthur Wellesley, Duke of Wellington, described soldiers as 'the scum of the earth … enlisted for drink', social outcasts that only 'punishment and repression' could hold in check (in Blanco 1965: 126).[2] Surveying eighteenth-century England, historian William Lecky

[2] Wellington used the phrase 'scum of the earth' during the Peninsular War in a letter addressed to Henry Bathurst on 2 July 1813—that is, after the storm of Badajoz (March–April 1812) when 'an orgy of drunkenness' occurred and Wellington 'lost control of part of his army for almost two days' (Keegan and Holmes 1985: 53). Wellington's view of soldiers was certainly more complex than the one we find in this letter. But it is interesting that, though he later spoke publicly about the positive qualities of the British soldier and regretted not being kinder to his men, in 1831, during a conversation with Philip Henry, Earl of Stanhope, he used once again the expression 'scum of the earth' to describe British troops (Coss 2005: 47).

observed that the army consisted of 'the dregs of her population' (in Conway 1987: 46), and Ian Roy (1977: 150) went even further when he defined the military as a 'criminal class'.[3]

The idea that soldiers deserve societal respect because they risk death in war is either an ancient belief or a recent one, as that was not the way common soldiers were perceived in early modern Europe. A soldier's life was relatively insignificant on or off the battlefield; what really mattered was his capacity to fight. Soldiers were viewed mainly as *faceless combatants* rather than beings endowed with individual personalities, and the army was conceived of and managed as an anonymous mass of human flesh. Hence, it is hardly surprising that European warfare in the early modern period was characterized by systematic disregard for soldiers' lives.

Reaction to soldier deaths was shaped largely by the social class of those killed in battle. Whatever the suffering of the families of the dead—something on which we have basically no knowledge—the loss of rank and file elicited no social response on the part of the political and military leadership. Their death in war was seen as a minor social occurrence, causing a mixture of indifference and pleasure. Some even took satisfaction in the death of their own poor combatants. Historian J. R. Hale (1985: 84) reminded us that Charles the Bold, duke of Burgundy, 'laughed off the slaughter of a third of his men when, after the debacle at Murten (1476), they were shot in the trees where they had taken refuge'. If there was any humane concern it was for the 'men of birth', not for commoners.[4] This class-conscious distinction is also apparent in Charles V's response to the loss of his soldiers through disease at the disastrous siege of Metz in 1552. Once the Duke of Alba had assured the Holy Roman Emperor that the dead were all 'poor soldiers', he declared that 'it makes no matter if they die, comparing them to caterpillars and grasshoppers which eat the buds of the earth ... and therefore no great harm if they died' (Parey 1825: 62). Human life was worth very little, and indifference to death in war was the norm. Thus, preserving the life of commoners was not a guiding principle in military decision-making and tactics.

This does not amount to saying that military commanders were uninterested in the lives of their combatants. For one thing, military advantage has always resulted from a combination of range, speed, and also protection from enemy

[3] That is not to suggest that everyone considered soldiers as a criminal class. For example, in *Lettera in difesa delli soldati stimati barbari*, the Italian general and geographer Luigi Federico Marsili criticized the idea that soldiers were 'barbarians' and described them as 'God's servants' (see Gherardi (2014: 303–4)).

[4] Charles the Bold fell in the battle of Nancy on 5 January 1477. As a sort of retribution for his disregard of soldiers, his body was found so horrendously mutilated by wounds and wild animals that only his personal physician could identify the remains.

attacks (Howard 1976: 3). In order to reduce casualties, commanders tried to cover and hide their troops by exploiting the advantages of the terrain and the topography of the battlefield.[5] And, as weapons became more and more lethal, armies gradually modified their tactics in order to minimize the killing potential of enemy weaponry. For example, on 13 September 1759 during the Seven Years' War, General James Wolfe first instructed his troops to lie down in order to save them from French skirmishers, and then ordered those same troops to rise and assault the French forces led by the Marquis de Montcalm (Lynn 2005: 187). Likewise, in order to avoid the direct line of French fire at Waterloo, Wellington's troops either used the reverse slope or lay flat on the ground (Browning 2002: 42).

Doubtless, protection from enemy fire through cover and armour was important in early modern Europe. Yet, commanders wantonly exposed their men to slaughter if military necessity required.[6] Although no reliable or precise figures are available on battle deaths, casualties appear to have been rather high. According to T. N. Dupuy (1984: 170), the rate of the killed and wounded in proportion to the soldiers employed in linear tactics during the Thirty Years' War was 19 per cent for the victors and 28 per cent for the losers. But the rate could be higher, as reported of Frederick the Great's armies at Zondorf (1758) and Kundersdorf (1759), where 38 and 48 per cent of men, respectively, are alleged to have been killed (Showalter and Astore 2007: 81, 91). Moreover, these estimates might underestimate the number of the war dead, not only because—unlike officer losses—casualties among enlisted men were rarely recorded (Clodfelter 2008: 4), but also because many of the wounded died in the days and weeks following battle as their injuries could not be properly tended. When casualty lists were compiled, they were not intended to reckon the amount of human lives lost in war, but rather to assess the relative strength of the army, which further illustrates how cheaply a soldier's life was sold in that age.

Tactical Changes, Unchanged Bodies

Battlefield tactics were characterized by widespread disregard for human life in war. Tactical innovations in the early modern era did not rate the protection

[5] In several passages of *The Art of War*, Machiavelli (2001) recommended soldiers should take cover from artillery fire.

[6] The possibility of seeking cover also depends on the type of weapon a soldier employed. Until the needle gun was introduced in the mid-1860s, riflemen could not fire and load their guns while lying on the ground (Neiberg 2006: 55).

of human life in combat zones as a primary concern. The transition from squares to line, one of the main tactical improvements in that period, was largely meant to exploit the deadly impact of firearms, regardless of the combatants' safety. The significance of this tactical change for the present research is worth pondering further.

In the late sixteenth century, the duke of Alba added to each *tercio* (that is, the Spanish pikeman formation) two companies equipped entirely with firearms. In the following decades, squares of pikemen protected by a few rows of marksmen were gradually substituted with long firing lines of musketeers defended by a few pikemen (Parker 2005: 154). While firearms were initially employed to ward off a pike charge, pikes were later used to protect firearms (Howard 1976: 56). Such a practice developed still further when William Louis of Nassau (Parker 2007) realized that 'rotating ranks of musketeers' could create a continuous, sustained stream of deadly fire capable of disrupting entire enemy ranks. Supported by field artillery, musket men fought in formations several lines deep. In this way close-order marching in line by infantry increased the firepower of the imprecise musket. As horsemen had no chance of survival against infantry firepower, they were mainly employed as shock troops.

In the 1620s, with improvements in the reloading speed of muskets, Gustavus Adolphus could achieve a constant volley of fire from his soldiers with only six ranks rather than the ten required in the Dutch forces.[7] After the battle of Breitenfeld in 1631, all major armies adopted linear formations (Parker 2005: 156–60), which became increasingly thinner and longer (Lynn 2005: 169) until they developed into the three-rank lines of eighteenth-century battlefields (Howard 1976: 60–1). This tactical innovation was completed by the introduction of the ring bayonet on top of the musket and later by the more effective socket bayonet, which eliminated the pike from European battlefields. Once enemy formations were broken by the firewall of infantry and artillery, the bayonet charge was to be used as a final thrust into the flesh of the opposing army.

Despite these major tactical breakthroughs, no change occurred in the attitudes towards soldier deaths. In this type of linear, flat warfare, discipline and drill were the key to success, because soldiers had to be prepared, in the words of historian Jeremy Black (1999: 18), 'to remain whole in the face of death, regardless of casualties'. Indeed, men in linear formations were quite exposed to

[7] Gustavus Adolphus was instructed by Jacob De la Gardie, who was a pupil of Maurice of Nassau (Howard 1976: 57). On Gustavus Adolphus's military campaigns and tactical innovations, see Roberts (1992).

the mortal danger of enemy fire. As Keegan (2004: 179) put it bluntly, the line infantryman had the 'unspectacular duty of standing to be shot at'. Drill and training were designed not only to teach soldiers tactical formations and how to handle weapons, but also to transform them from emotional human beings into cold-blooded automata with no natural fear of being shot to death, which was an essential requirement at moments of great danger when—given the inaccuracy of muskets—they were told not to fire first (Lynn 2011: 93–4). In this type of warfare, the personal bravery of the individual soldier was secondary to preserving order among the mass.

Not coincidentally, a revolution in drill came about in the seventeenth century (McNeill 1982: 126–43).[8] Since inflexible discipline was required to withstand devastating volleys at close quarters, the importance of robot-like obedience to orders could not be underestimated. Moreover, the typical smoke, noise, and confusion of the battlefield left soldiers little time to think what to do. Successful combat in linear formation required soldiers to stand calmly at point-blank range and fire mechanically into the enemy line, waiting for a shot that could literally blow them apart. Drill was aimed at depriving a man of his natural instinct to survive in order to rebuild him as a totally subservient component. The end product of drill was the *line soldier*, who was expected to behave like a puppet. Men under arms were made to act out automatic movements turning them into tiny parts of a deadly mechanism, which was much larger and more important than individual welfare. John Keegan (1993: 9) gives a detailed description of fighting conditions in early nineteenth-century warfare: 'Men stood silent and inert in rows to be slaughtered, often for hours at a time; at Borodino the infantry of Ostermann-Tolstoi's corps are reported to have stood under point-blank artillery fire for two hours, "during which the only movement was the stirring in the lines caused by falling bodies".' Although Borodino was no ordinary battle, the nature and dynamics of the action did not differ significantly from the warfare of the time, in which protection of armed forces was secondary to inflicting heavy losses on the enemy.

The transition from a solid mass of infantry to line order was determined largely by the gunpowder revolution, since infantry when compacted could be destroyed by cannonballs. The linear deployment of soldiers was a tactical change aimed not only at massing fire forward, but also at minimizing the lethality of incoming shots. However, spreading out the troops on the battlefield was not a result of humane concern for soldiers' life, but mere tactical

[8] On the importance of drill (and dance) for explaining human evolution, see McNeill (1995).

expediency. Furthermore, as John Lynn (2011: 94) pointed out, warfare in the seventeenth and eighteenth centuries was 'not so much' about 'inflicting casualties as demonstrating that one could absorb enemy fire and continue to fight and advance without breaking or flinching'.[9]

Another major tactical innovation, passing from linear formation to columns of attack based on rapid, massive charges into enemy lines—typical of Napoleonic warfare—was adopted in that period, even though it led to a further increase in casualties (O'Connell 1989: 175). For there was a rise in French casualty rates from 9 per cent for victors and 16 per cent for losers to 15 and 20 per cent of the effective combat forces (Dupuy 1984: 170), which could be sustained only thanks to the increased manpower made available by the Revolution.[10] The use of infantry columns of attack, *l'ordre profond*, was not an invention of revolutionary France. It had been employed during the War of the Austrian Succession (1740–8), when French columns were 'shredded to pieces' by the fire of the enemy line (Howard 1976: 79). That is why it was temporarily abandoned in the following decades. However, by supplying reliable, motivated men in unprecedented numbers, revolutionary France was able to sacrifice amounts of lives that the socially conservative *ancien régime* could not afford. Thanks to the decree introducing *levée en masse* on 23 August 1793, a new type of virtuous citizen-soldier was forged and, for the first time in history, large swaths of citizens were called to arms in order to defend the nation.

Nevertheless, political and ideological transformation did nothing to modify attitudes to men fighting on the battlefield. The French Revolution boosted the capacity of the state to recruit combatants, but the meaning attached to soldier deaths remained largely unaltered. The French soldier was rhetorically treated as a national hero, but that acknowledgement was principally meant to justify the massive losses incurred by the French army and Napoleon's *Grande armée*. One should remember that, of the 533,000 men Napoleon employed to invade Russia, only 44,000 returned to France. Also, discriminatory attitudes towards life and death persisted within the army, as attested during the Austrian campaign of 1809, when the wounded of the Imperial Guard were immediately evacuated, while the other French casualties were still awaiting

[9] Even the oblique order employed by Frederick the Great in order to concentrate firepower on one wing and, in doing so, produce a decisive attack was not meant to reduce casualties, but rather to obtain a tactical advantage regardless of human losses. Sustaining heavy casualties was not a major problem if it made sense strategically.

[10] The higher casualty rates produced by columns derived from the simple fact that this more compact formation represented an easier target for enemy fire than the line (Browning 2002: 24). Overall, including the dead through illness, Napoleon's armies had a mortality rate of 40% (Delmas 1992: 329).

evacuation three days after the battle (Richardson 1985: 13). In short, apart from the purely strategic viewpoint, there is no evidence that the problem of soldiers' expendability played a role in the evolution of military tactics or in the conduct of military campaigns in early modern Europe, even in the decades following the French Revolution, which is generally considered to be the beginning of the modern period.

The Wounded

While it is true that in the seventeenth and eighteenth centuries governments spent money and time on building and quartering soldiers in barracks and on providing them with medical services, preoccupation with soldiers' welfare and health still lay in their military potential on the battlefield, since more soldiers meant more military power.[11] The main goal of medical treatment in this period was to return the largest number of men to the killing fields as quickly as possible so that the butchery could be carried on until victory was achieved. This was, for example, the function of what historian Geoffrey Parker (2004: 141) considers the first permanent military hospital on European soil, which was opened in 1585 at Mechelen by the Spanish Army of Flanders. But, not surprisingly, medical aid to the wounded was forbidden until battle was over.

Patching up men only to enable them to kill once again was also what the French had in mind when they established mobile ambulance hospitals. By the end of the eighteenth-century, France had 1,200 doctors for 200,000 soldiers (Browning 2002: 14), but medical care remained a low priority, especially when it served no military purpose. Moreover, linear warfare, with infantry formations scattered across the battlefield, made location of the wounded even harder. Thus, the wounded were often left untended on the battlefield, crying for aid, water, and—as likely as not—a quick death. The fate of the wounded depended largely on the benevolence of local inhabitants and, of course, the victors, who frequently massacred men who had fallen on the field.[12]

[11] Most countries housed their troops in inns, sheds, and shacks throughout towns or in farmhouses, whereas France used barracks; by 1775 over 300 French towns possessed military quarters fitted out for 200,000 men (Wilson 1999: 75–6).

[12] Captured nobles were usually excluded from the practice of butchering prisoners and the wounded, being spared for ransom.

Ambulant hospitals, ambulances, and later flying ambulances, introduced by Napoleon on the heels of Dominique-Jean Larrey's invention, greatly increased the capacity for locating, evacuating, and treating casualties (Rostker 2013: 24). Other developments in military medical services in the nineteenth century improved both the delivery of medical care and the chances of survival of the injured, but, even then, medical treatment was given first to the lightly rather than the severely wounded, in order to bring them back onto the battlefield (Neiberg 2006: 85). When it came to a tradeoff between soldiers and materiel, there is ample evidence that commanders went for the latter. During the Peninsular War, for example, flying ambulances were not employed for the treatment of the wounded, because Wellington refused to give priority to the tending of soldiers. In a situation of short supplies, priority was given not to medical treatment (Browning 2002: 47), but rather to military equipment. Soldiers were valuable as fighting tools, not as human beings.[13]

Finally, it is worth remembering that, until the twentieth century, the ravages of disease and sickness killed far more soldiers than actual combat on the battlefield (Gabriel 2013: 58). For soldiering involved many lethal dangers besides combat. Poor camp hygiene and food, filth and crowded billets would weaken soldiers and expose them to numerous pathogens. A variety of communicable diseases would spread throughout the army camp, where close quarters and minimal facilities were more harmful than warfare. Even during the brutal Thirty Years' War, plague and war-related food shortages, rather than actual fighting, were the main causes of death (Parker 1984: 211). It is no overstatement to say that European armies were marching health calamities. The most lethal killers in the early modern period were typhoid, fever, scurvy, dysentery, or tuberculosis, not guns and bullets. And, again, soldiers were doomed, not just because of lack of medical knowledge, but also because the officer corps cared very little about their men.[14] Studies on mortality rates in Napoleonic armies (Delmas 1992: 328) show that the likelihood of dying from illness was more than twice as high for ordinary soldiers as it was for officers. Medical 'care on the battlefield', concludes Richard Gabriel (2013: 54), 'was still mostly limited to kings and nobles ... The idea that medical aid should be extended to the common soldier had yet to take root.'

[13] It should also be noted that, though some of Wellington's Peninsular veterans were admitted to the Royal Chelsea Hospital, most were left to their own devices (Keegan and Holmes 1985: 162).

[14] Before the mass production of quinine to cure malaria, the death rate of soldiers garrisoned in colonial territories could be appalling. Quite correctly, Charles E. Callwell defined colonial wars as 'war against nature' (in Wesseling 2013: 35).

Limited Warfare in the Eighteenth Century?

Arguing that soldier deaths in a sort of organized mass slaughter were of little concern to statesmen and generals alike does not equate to suggesting that commanders were bloodthirsty or eager for butchery. Europe after the Thirty Years' War, especially between 1715 and 1789, is known as an area and an epoch of limited war in which large-scale killing was generally avoided and battles could be won through manoeuvering and with little or no fighting (Lynn 2005: 178).[15] According to Pascal Vennesson (2011: 252), the regard for troop protection, typical of the old regime, is similar to the limits and restrictions imposed on the conduct of war in the post-cold-war era. In particular, Vennesson suggests that eighteenth-century limited warfare is an early case of casualty aversion.

Although European armies in the eighteenth century focused mainly on siegecraft, manoeuvre, marches, and supply and less on infantry battles, it was not humane concern for soldiers that influenced how they were employed or not employed in war. The limited nature of warfare and the preference for outmanoeuvering one's foe rather than forcing him to battle—typical of the second half of the eighteenth century—appears to be explained not by normative reasons, but rather by two principal constraints: good, well-trained professional soldiers were a scarce and expensive resource and the large armies of the time were formed of 'many unsuitable, untrained and untrustworthy men' (Hale 1985: 64), which made battle a risky gamble.

In regard to the first reason, Clausewitz (1984: 590) observed that often, if 'the army was pulverized', the commander 'could not raise another, and behind the army there was nothing'. In early modern Europe, good soldiers were a commodity in short supply, difficult to replace. Accordingly, men were to be used with caution. Statesmen and generals did not want to have their expensive investment in recruiting and training destroyed on an unlucky day of fighting (Woloch 1982: 56). Hence, commanders attempted to salvage their costly manpower unless they believed there was a high probability of success.

As for the second reason, it should be stressed that desertion was a common and pervasive phenomenon in early modern Europe, despite all attempts to prevent it. The rate of desertion was different for each country and could range from one-sixth of the British army in Ireland in the 1780s (in Best 1998: 33)

[15] The War of the Spanish Succession came to an end in 1714. In the following years, with the death of Louis XIV, France's continental ambitions were replaced by the more moderate foreign policy of Louis XV.

to 40 or even 50 per cent. Defection was especially high when soldiers were not paid. As the saying went, *pas d'argent, pas des Suisse* (no money, no Swiss): when unpaid or driven too hard, soldiers would desert and even mutiny.[16] Not only did mercenaries such as the *Landsknecht*—a German replica of the Swiss pikemen—regard themselves and often act as free men, but most soldiers were seen as potential deserters (Lynn 2005: 180). Mutinies for lack of pay were a common threat. Not even Spain, the global power of the sixteenth century, was able to pay its soldiers on a regular basis. On this point, Geoffrey Parker (2004: 157, 177, 253–6) reminds us that the Spanish infantry in the Netherlands 'received hardly any pay' between October 1570 and May 1572. The inadequate system of payment and supply was one of the factors that brought about forty-five mutinies in the Spanish Army of Flanders between 1573 and 1607; and many of them lasted for a year or more.

The situation did not significantly improve in the eighteenth century, especially at harvest time, when men could find employment as agricultural labourers in the countryside. Military historian Hew Strachan (1983: 9) provides these striking figures for two of the most important conflicts of the period: 'One in four Frenchmen deserted from the army in the War of the Spanish Succession, and from 1717 to 1728 there were 8,500 deserters for every 20,000 men in the Saxon infantry. In the Seven Years' War, 80,000 men absconded from the Prussian army, 70,000 from the French, and 62,000 from the Austrian.'

In addition to the risk of desertion, logistical factors also set limits to the intensity of warfare, especially problems of supplies and transportation connected with the movement of large armies (Creveld 2004). Strachan (1983: 11–15) concludes that war in the eighteenth century was

> limited by its means, but this was in spite of practitioners, not because of them … means were not limited if they were consonant with the objectives … Attitudes to war and to battle were therefore no more humanitarian than in any other age. If the effects of battle were limited, this was due to the practical constraints of the times and not to design.

In other words, limited warfare was largely a result of two associated circumstances: the limited availability and costliness of good soldiers, as well as the unreliability of traditional old-regime combatants.

[16] In a famous passage of *The Prince*, Machiavelli (1998: 48–9) writes that mercenaries 'have no love nor cause to keep them in the field other than a small stipend, which is not sufficient to make them want to die for you'.

Furthermore, no remarkable difference existed between attitudes towards conscripts, on the one hand, and volunteers and mercenaries, on the other.[17] Although foreign troops could be wasted without the agricultural manpower of the nation being depleted (Woloch 1982: 52) and conscripts were considered relatively more trustworthy than mercenaries, these different categories of combatants were heedlessly sacrificed to the 'god of war'. Soldiers were nothing but things 'to be dressed up, marched around, and shot' (O'Connell 1989: 154) in the deadly tournament of war. In early modern Europe, the soldier had no genuine personal or social identity; he was merely a physical corporality whose life could be expended according to the strategy and tactics of commanders.

Another peculiarity of warfare that shows how lowly soldiers' lives were rated at that time is the type of punishment they received when they refused or failed to face the enemy in battle: the penalty for cowardice was often death. Not only did Frederick the Great decree that non-commissioned officers were to kill soldiers who turned round in the ranks under fire,[18] but General James Wolfe himself issued the following brutal order in 1755: 'a soldier who quits his rank, or offers to flag, is instantly to be put to death by the officer who commands the platoon, or by the officer or sergeant in the rear of the platoon; a soldier does not deserve to live who won't fight for his king and country' (in Hall 2016: 119). 'Command well, pay well and hang well': that is what a good general, according to Lord Ralph Hopton, ought to do if he wanted to be successful in war (in Carlton 2003: 81). Soldiers were left no choice but to fight or to die, and there was often no real difference between the two.

The fact that military casualites were considered socially and morally unproblematic is in striking contrast to the current view that protecting the lives of western troops is a necessary requirement for their deployment in combat zones. In early modern Europe, no tacit or codified contract existed between

[17] Forms of partial conscription existed. For example, with the Ordinance of Valladolid in 1494, conscription was introduced in Spain for one man in twelve between the ages of 20 and 45 (Howard 1976: 32). Sweden, Prussia, France, Denmark, and Austria also used some forms of partial conscription. For example, from 1625 to 1630 about '50000 men were conscripted in Sweden, according to a rota system which ensured that every village provided a certain number of men—by no means all adult men had to serve' (Asch 1997: 164). Conscription was imposed in France in 1688 among unmarried peasants, in Denmark in 1701, in Spain in 1704, in Russia in 1705, and in Austria and Bohemia in 1771 (Black 1999: 13). According to Charles Tilly (1992: 80–4), European states increasingly resorted to conscription, because nationals were cheaper and considered more loyal than foreign mercenaries.

[18] Historian Isser Woloch (1982: 57) reminded us that 'troops were supposed to fear their officers more than the enemy ... Prussian non-commissioned officers were equipped with spontoons, long implements designed less to wound the enemy than to prod their own men.'

the state and its servicemen,[19] by which soldiers put their lives at risk in exchange for the assurance that the state would do everything possible to ensure that they were not going to be killed. Neither was the ancient and current rule *leave no man behind* followed at that time (Samet 2005; Wong 2005). This policy—a feature of current military practice, which was seen at work, for instance, during the rescue operation organized to save USAF pilot Scott O'Grady after his F-16 had been shot down by Serbian anti-aircraft artillery in June 1995—was non-existent in early modern Europe. If a man was missing in action or wounded, his officers did not feel obliged to organize any rescue mission. Commanders were largely uninterested in the fate of the individual soldier, a mere instrument of war.

In early modern Europe, the only important criticism as to the use of force in war was about the way soldiers began to kill with the introduction of gunpowder weapons, and not about dying and killing *per se*. Firearms were profoundly despised by the advocates of chivalric ideals, vividly exposed in two of the most influential works of early modern European literature: in Ariosto's *Orlando furioso* (1532) and, with great irony, in Cervantes's *Don Quixote* (1605–15). Both Orlando and Don Quixote were traditional knights, disgusted with gunpowder and pistols, because these weapons were considered fit for cold-blooded murderers, not for truly honourable warriors (see Arnold 1999: 38).[20] For one thing, by enabling commoners to kill noble men-at-arms—or, as Blaise de Monluc put it, 'valiant men' slain by 'the most pitiful fellows and the greatest cowards' (in Cassidy 2003: 387)—firearms erased the distinction between these two social classes in favour of the former. In a battle in which death comes from guns, the traditional skill and training of warriors counted for little. Moreover, being difficult to reload while on horseback, firearms jeopardized the chivalric code of fighting typical of the war of knights (Creveld 1989: 88). Although killing from a distance was not introduced with the gunpowder revolution, the latter put an end to the ideal of *chevaleresque* warfare.[21] And that was why hand-gunners were not only despised in early modern Europe,

[19] For a formal contract between the nation and its soldiers, see the British Military Covenant, formally codified in 2000, which explicitly states that the country has a 'duty of care' to its armed forces. On the British Military Covenant, see McCartney (2010).

[20] The following words against long-range firearms are proffered by *Don Quixote*: 'Blessed be those happy ages, that were strangers to the dreadful fury of these devilish instruments ... which is the cause that very often a cowardly base hand takes away the life of the bravest gentlemen, and that in the midst of that vigour and resolution which animates and inflames the bold' (in Silver 2002: 61). On the arquebus as an evil weapon subverting chivalrous custom, see *Orlando furioso*, canto IX (28–30).

[21] Paul Bingham (1999) argued that the ability to kill at a distance through projectile weapons gave human beings a comparative advantage over animals and was the key to our survival and expansion.

but also underwent especially cruel treatment when captured, such as amputation of hands or fingers (Hale 1985: 95), a harsh measure that would be reserved for snipers in the twentieth century.

What caused knights to be upset was not the deadly power of the weapons, but the way in which they killed, the fact that they did not kill 'well', with dignity. Gunpowder marked the end of the chivalrous tradition in western warfare, which linked death on the battlefield with face-to-face combat. In other words, the traditional code of chivalric warfare was destroyed by the new weapons, especially handguns, which allowed a soldier to kill his enemy without meeting him in open combat.

Finally, in spite of the stereotype that describes eighteenth-century warfare as restrained, war in that age was far from being limited in terms of casualties. For example, at the battle of Malplaquet (1709) during the War of Spanish Succession, 36,000 men were killed or wounded on a single day. More importantly, the figures available suggest that the eighteenth-century ratio of combat deaths to the total population of Europe was seven times as high as the rate recorded in the nineteenth century (Showalter and Astore 2007: 67).

If there were any limits or instances of 'fair play', they were reserved mainly for the world of gentlemen. The 'commendable practices' of warfare and the 'high degree of courtesy' exhibited in war, mentioned in a variety of distinguished sources—such as Emmerich de Vattel's *The Right of Peoples* (1785)—were limited in fact to the officer corps. Not only was it common for officer prisoners to have dinner in the mess of the enemy (Neiberg 2006: 89) and to keep their personal weapons when captured, but members of the officer corps had privileged treatment in battle as well: Wellington's order not to shoot Napoleon at Waterloo is an example of the niceties reserved for fellow generals, from which common soldiers were totally excluded.[22] Such niceties, it should be noted, were not reserved for one's own rank and file. For instance, after the battle of Waterloo, the wounded of Wellington's army remained untended as long as eleven days after the clash (Gabriel 2013: 150).

Besides the immense bloodshed in eighteenth-century battlefields, there is a great difference between the aversion to casualties of the Westphalian period and the aversion typical of the post-Second World War era: in early modern Europe, the focus was on the army at large and on its military capacities, not on the lives of individual men. Unlike the present era, no casualty sensitivity

[22] The privileged treatment reserved for officers, however, did not always translate into safety during battle. Losses among the officer corps could be higher in proportion to those among commoners, especially in sea battles and siege warfare (Clodfelter 2008: 4).

existed before the twentieth century, while soldiers—it is worth repeating—were highly expendable if the exigencies of war so required. War was a 'royal sport',[23] and expending soldiers was one of its basic components. Resorting to the commonly used chess analogy, one might argue that the survival of pawns was not important in itself, but only in order to succeed in the great game of war. Death in war was perceived as part of an accepted order of things. As Jeremy Black (1999: 2) observed: 'Killing was not seen as unnatural, but was, instead, generally accepted as necessary, both for civil society, against crime, heresy and disorder, and in international relations.' Black's description of the major role violence played in early modern Europe needs just a small amendment: not only was killing seen as something natural and unproblematic; dying in war was viewed in similar terms.

Like Animal Carcasses

In the summer of 1750, during a visit to the battlefields of Flanders, Voltaire noticed the absence of markers of burial places and remarked that there 'was nothing there anymore; everything was covered by the most beautiful wheat in the world; the Flemings danced as if nothing had ever happened' (in Troyansky 1987: 121). No indication of carnage was visible, no crosses, no monuments, no reminders of the conflict, nothing at all. This was not a time when the battlefield was a landscape of memory and remembrance; actually the scenery was remarkable only for its pastoral serenity. Not much was owed to the ranks when alive, and nothing, as Voltaire could tell after his journey to Flanders, when they were dead: not even a tomb.

Two centuries later, Edward Steere (1948: 149), engaged in research in the Historical Section of the US Office of the Quartermaster General, lamented that, although

> many glorious memorials have been erected in ancient and modern times to commemorate the fame of great statesmen and soldiers, it is a melancholy fact that only within the past hundred years has any government been willing or able to assume the obligation of identifying and burying in registered graves the remains of all who gave up their lives in war.

[23] About Britain, French historian Élie Halévy pointed out that war 'was like any other outdoor sport, only rougher and more dangerous' (in Blanco 1965: 128). Galileo Galilei seems to have been the first to describe war as a 'royal sport' (in Hale 1985, 29).

Before the nineteenth century, as a matter of fact, the corpses of the war dead were not buried in individual, registered graves in military or civilian cemeteries. They were unceremoniously interred either in the proximity of the battlefield in individual unmarked burials or, most often, in shallow mass graves. Corpses were thrown into a pit and tumbled together into complete oblivion.

At best, burial of the war dead was regarded as a problem of sanitation, rather than a social requirement. There were no military cemeteries for them, but only excavated holes where their corpses were simply left to rot without causing miasma.[24] Soldiers' remains were reduced to refuse and disposed of without ceremony in anonymous common graves.[25] Their burial was a mere form of interment without the usually related funerary rites. If the victorious army did not take charge of burial operations, then the local population was supposed to provide the service. In such cases, by the time soldiers were buried, their bodies had already rotted, and in the following days and weeks one might see pieces of corpses scattered across the field. As soldiers were regarded as 'food for worms', battlefields were receptacles of putrefying human flesh.

The disrespect for the dead was such that not only were bodies regularly robbed of their personal effects, but also physical parts like teeth were removed and sold on the market (Summers 2010: 11). At the time it was considered perfectly normal to pick up a tooth on the battlefield of Waterloo and embed it into the mouth of another man, as happened in January 1816 to the war correspondent and diarist Henry Crabb Robinson (Semmel 2000: 9). And it was also perfectly normal that after the battle of Waterloo 'English contractors collected the bones of the dead from both sides, ground them up, and sold them to fertilize English gardens' (Ignatieff, in Gray 2003: 218).

Even the exact location of mass burials was of little importance, and in most cases not even marked by an inscription. In the early modern period, the issue of where and how to bury the war dead and of honouring their graves was not an issue at all. Most of the fallen were buried in pits, and their death remained

[24] Despite the common view, the smell from decomposing bodies, though unpleasant, does not usually pose a health risk to survivors. Battle dead rarely caused epidemics. Having being killed by weapons, soldiers' corpses do not usually have any threatening contagious agents. Moreover, most infectious organisms do not survive beyond forty-eight hours in a corpse (Morgan et al. 2006: 5). Owing to epidemics of diseases such as cholera, typhoid, and malaria, the surviving soldiers—rather than the dead—were more likely to spread illnesses and pose a real lethal threat to their comrades and the civilian population. On the myth of epidemics caused by dead bodies, see Ville de Goyet (2004).

[25] On a page of *If this is a Man*, speaking of life in the Nazi camp where the inmates had their clothes, shoes, hair, and names immediately taken away, Primo Levi (1959: 21) noted that only a human being deserves to have a name. Levi observed that the ultimate process of dehumanization was precisely the stripping of the name.

unrecorded as a result. At any rate, providing individual burials would have been largely meaningless, since there was no systematic way to identify the dead.[26] After all, the impossibility or inability to locate and identify the bodies of the fallen in early modern Europe does not seem to have created any special anguish or problems at the societal level.

Even the heroes of the American Revolution—men who had fought to establish the first modern democracy in the new world—did not deserve the privilege of individualized burials. During that conflict, as in the early modern period, most of the war dead were buried in trenches and mass graves and, as historian James I. Robertson (1988: 225) explained, that 'gruesome and nauseous job was done with haste rather than reverence'. Even outside the battlefield, soldier deaths remained unnoted. For instance, at Valley Forge, the site of the camp of the American Continental Army over the harsh winter of 1777–8, virtually none of the 3,000 soldiers who had died of starvation and disease were buried in individualized, marked graves.[27]

Even in the first half of the nineteenth century, mass graves were still the most common type of burial employed in war. In countless instances, the war dead met with no orderly dignified disposal. A traveller who walked across the field of Austerlitz (1805) a few weeks after the battle provides a typical description of how the bodies of the war dead were dealt with in that epoch:

> Traces of the slaughter were still to be seen all over the field. The many freshly-turned piles of earth betrayed the sites of the graves. We repeatedly had to hasten past spots where inadequately buried bodies of horses had been worked over by the ravens and crowns, and were spreading intolerable stench. Worse still, you could see the severed limbs of dead or maimed soldiers, together with half-eaten skulls, bones and ribs.
>
> (in Keegan and Holmes 1985: 160)

[26] From the sixteenth century on, the necessity to distinguish between friend and foe as well as among different units led to the use and later production of uniform markings. At the battle of Pavia, in order to make identification possible on the dark night of 25 February 1525, soldiers of the imperial army were ordered to wear their shirts over their armour (Showalter and Astore 2007: p. xxvi). At that time there were no standardized uniforms, only loose, rough markings. It was only in the second half of the seventeenth century, when armour was abandoned and soldiers were organized in permanent regiments, that real uniforms became the regular outfit (Hale 1985: 163–4; Parker 2005: 153). According to Roberts (1992: 237), Gustavus Adolphus was the first military commander in early modern Europe to dress his troops in similar clothes, and France the first nation to establish uniform clothing in the 1670s. However, and interestingly for this topic, no names were marked on uniforms or on the bodies of individual soldiers. Hence, there was no way to identify the cadavers. Besides, while calling a truce in order to collect and bury the dead was a common military custom in ancient times, that was comparatively rare in early modern Europe.

[27] At Valley Forge, according to Thomas Laqueur (1994: 158), only the burial of a lieutenant from Rhode Island was marked.

After the battle of Austerlitz, the 18,000 victims were interred in 25 mass graves (Rigeade 2008); likewise after the epic clash of Waterloo, the English novelist William Makepeace Thackeray lamented in his *Little Travels and Roadside Sketches* how common soldiers were 'shovelled into a hole ... and so forgotten' (in Blunden 2003: p. xx). Chapter 3 will show that concern for proper burials of dead soldiers would not develop until the American Civil War, but even during that conflict corpses were often left to decompose unburied. Although various actors involved in the conflict—the Federal government, the War Department, soldiers, and citizens alike—tried as best as they could to provide decent disposal for the dead by interring them in suitable spots adjoining the battlefield (Grant 2005: 512), historian Drew Gilpin Faust (2008: 69) reports that for weeks after the battle of Gettysburg approximately six million pounds of human and animal flesh were left putrefying in the summer heat; and, for weeks afterward, the local population had to use pennyroyal and peppermint oil to counteract the stench.

The main alternative to mass graves was not individual burial, but rather destruction through fire. Such a practice, which belongs to a long history of emergency situations such as epidemics, natural disasters, and—obviously enough—war, was still used for sanitary reasons in the early modern period and even in the second half of the nineteenth century. After the battle of Sedan (1870), for instance, the Belgian government dispatched Colonel Creteur to cremate the decaying cadavers of German soldiers who were covered only by a thin layer of soil (Erichsen 1887: 136–8).[28]

As a result of mass burials and, very rarely, mass cremation, most men from the ranks perished anonymously as if they had never been alive, which means that they were devoid of a proper social body. The absence of military cemeteries and individualized graves implies that marking the location of the physical remains of the war dead was regarded as utterly unimportant. The only concern in burial practices was 'to sort our noble from our common men', as the French asked Henry V after the battle of Agincourt (1415) (in Blunden 2003: p. xix). Such a discriminatory practice was still in use in the eighteenth century, as can be extrapolated, for example, from the orders of Lieutenant-Colonel Hans Friedrich von Leming, who in 1726 requested his higher command both to prevent 'the dead from being plundered because

[28] Although in the ancient world cremation was often employed, this form of disposal was far from 'speedy and efficient' (Noy 2000: 186). Owing to logistical problems and the large amounts of fuel (mainly wood) required for the incineration of bodies, destruction by fire remained a highly inefficient way to dispose of corpses until the early twentieth century, when improvements in ovens turned cremation into a cost-effective process. Modern cremation was reborn in Italy in the last quarter of the nineteenth century (Colombo 2017).

wounded officers are sometimes killed so that they may be stripped of their possessions' and to 'inform the enemy that the dead are to be buried so that they may come and look for high-ranking individuals who have been declared missing' (in Capdevila and Voldman 2006: 39).

Only officers, who belonged to the upper class, had individualized tombs and graves.[29] In a strictly hierarchical society, some individuals mattered more than others, and some bodies were more bodies than others. Officers maintained in death the privileges of wealth and rank they possessed in life. Thus, it is not at all surprising that the first military category to be granted single graves were those 'men of birth' (generals and officers) who had been the only real concern of their commanders when alive.[30]

Difference in attitudes towards the death of a commoner and that of a military commander was also strikingly visible during funerary rites. There was a sharp contrast between the care displayed at the death of a member of the officer corps and the brutal disregard for the remains of a commonplace soldier. Ceremonies for generals especially were incredible displays of splendour and riches. Suffice it to mention one of the most astounding military funerals ever organized, that of the duke of Wellington in 1852. An estimated crowd of 500,000 people saluted the coffin of the Iron Duke, in a spectacular ritual of public grief in which the English people gave the last salute to the 'saviour of Europe' and 'the greatest of living men', as Wellington was known at the time of his death.

That was at a time when the only bodies that mattered were those of the most illustrious and politically important dead. Commoners, instead, were social outcasts, branded even in death as socially worthless individuals, their corpses disposed of not unlike animal carcasses. Soldiers' decomposing bodies were not perceived differently from what the fictional De Sade said to Marat in Peter Weiss's play: compost that drowns in the total indifference not just of nature

[29] Only in emergency periods, such as when a plague epidemic broke out, did mass graves became the norm for the affluent part of society too. For example, in his account of the great plague that descended upon Athens, Thucydides (1998: ii. 52) describes how the epidemic simplified burial rites and diminished the traditional care over the disposal of bodies.

[30] Most officers came from the nobility and the landed gentry, usually the second and third sons of the aristocracy, just because they could purchase commissions. For example, almost 90% of Spanish officers after 1776 were drawn from the nobility (Black 1999: 4). Likewise, by 1786 only one-tenth of Prussian officers 'were commoners, and among the higher officers only 22 out of 711' (Woloch 1982: 55). Of the approximately 10,000 officers in the French army in the decade that preceded the Revolution, only 46 had been promoted from the ranks (Neiberg 2006: 8). Similar numbers can be found in the other European armies of the time (Tallet 1992: 85–92). Even in liberal Britain, the purchase of commissions disappeared as late as 1871. In the United States, where aristocracy of birth was replaced by the aristocracy of money, only wealthy families could finance the return of their loved ones during and in the aftermath of armed conflicts like the Revolutionary War, the war of 1812, and the Seminole Wars (Sledge 2005: 32).

but also of society, which was largely unmoved by their death—an attitude that would be considered ruthless and socially unacceptable by present-day standards.[31]

While death has often been portrayed as the leveller of human beings because it kills everyone irrespective of his or her social status (May 1972: 476), the way in which the corpses of soldiers were dealt with was significantly unequal in comparison to the treatment given to the affluent part of society. Common soldiers, largely recruited either from the countryside (Hale 1985: 124–5) or from urban labouring masses, were perceived as the scum of the earth and treated accordingly: as an anonymous, amorphous collective. While funerals and burial rites had the function of repairing the social damage caused by death, there was no need for commoners to re-establish their social fabric.

That the powerless and the lower orders were written out of the historical record is not surprising, but it should be emphasized that such disrespectful treatment of corpses—at least by today's standards—was unheard of even in some ancient societies. For example, although in Republican Rome the lower classes were often buried in mass graves (*puticuli*), in the second and third centuries AD the humble, the poor, and also loyal slaves employed in a household were 'no longer always buried in remote places close to refuse dumps' (Ariès 1985: 4–5).[32] Moreover, both in Athens and in Sparta, honour was paid to the remains of soldiers killed in battle, who were commemorated by stone tablets with their names engraved (British Academy 2012: 119).[33] From this perspective, we might conclude that the social status of the common soldier in early modern Europe reached one of its lowest levels in history—a status that in France was even lower than that of the poor after the *Decree of 23 Prairial,*

[31] I am referring to Weiss's play *Marat/De Sade*, especially to the following passage: '*Marat*: I read in your books De Sade … that the animating force of Nature is destruction and that our only instrument for measuring life is death. *De Sade*: Correct Marat. But man has given a false importance to death. Any animal plant or man who dies adds to Nature's compost heap becomes the manure without which nothing could grow nothing could be created. Death is simply part of the process. Every death even the cruellest death drowns in the total indifference of Nature. Nature herself would watch unmoved if we destroyed the entire human race (in Weiss 1970: 31–2). Although there was a divide between humanity and animality—soldiers' corpses were in fact generally buried—there is no doubt that in this period soldiers' remains were reduced to mere rotting flesh.

[32] Being slaves did not indicate much about their life and death in ancient Rome. Although all slaves were considered property under Roman law and did not have legal personhood, a great difference existed, for example, between domestic slaves who performed services in a household and slaves working in quarries and mines, whose life and death were especially miserable (Joshel 2010). In addition, domestic slaves had more chance of being buried in a decent way than the free poor, who lacked the financial resources required for decent disposal (Graham 2006).

[33] On the Athenian concern for the war dead and the practice of listing their names, generally on *stelai* above their common grave, see Loraux (2006) and Arrington (2014). Roman soldiers were generally disposed of collectively, via either cremation or interment (Hope 2018). Roman law and religion prescribed that the remains of the dead—including cremated remains—were to be buried.

year XII (12 June 1804), which made it compulsory for each civilian corpse to have a coffin and required that bodies buried in mass graves, including those of the pauper, be placed side by side rather than on top of each other. Therefore, when it came to mortuary practice, soldiers in the early modern period were treated as the poorest of the poor. Having had a miserable life and a miserable death, they were considered to deserve only a miserable disposal.

When Soldiers Were not Worthy of Memory

After describing how common soldiers were treated when alive and dead, let us now turn to how they were (not) remembered and commemorated. As we have seen, soldiers' bodies, dead or alive, were treated with indifference in the early modern period, but the preservation of their deeds did not rank high either. Memorials in the modern period commemorated war, especially a victorious conflict, and only very special casualties of war, primarily military and political commanders such as *condottieri*, generals, kings, and emperors. For example, the famous equestrian monuments by Donatello to Gattamelata in Padua and by Francesco Mochi to Alessandro Farnese, duke of Parma, in Piacenza commemorate military commanders, not commoners or ordinary soldiers. The construction of statues and memorials was to honour 'great men' and extend their social standing beyond natural death, but no sites of memory were built for ordinary people, whose deeds were not considered worth recording. As J. R. Hale (1985: 84) argued, in early modern Europe we 'are not to expect village war memorials, let alone tombs for the "unknown soldier"'.

Philippe Ariès (2009: 79) aptly argued that 'hero cults form the common denominator throughout western civilization in the late eighteenth and early nineteenth centuries'. Doubtless, the cult of great men is clearly visible at the Panthéon in Paris and in a city filled with monuments to revolutionary heroes such as Washington. However, celebration of the dead was restricted to illustrious ancestors, not to those who bore the major brunt of war. Even if the Panthéon, opened in 1791, was meant to commemorate merit and not birth (Ben-Amos 2000: ch.1, 'The French Revolution'), the memory of plain and humble soldiers was not to find a place in the mausoleum of the Revolution. Similarly, with few exceptions, the commemoration of the American Revolutionary War did not include the average soldier,[34] but rather great heroes

[34] One exception is the Revolutionary Monument in Lexington, erected in 1799 to commemorate the eight Minutemen killed by the British on 19 April 1795 (Linenthal 1991: 13). Although the monument is a traditionally shaped granite obelisk meant to commemorate the beginning of the Revolutionary

such as Washington, Joseph Warren, and Anthony Wayne.[35] Tellingly, in the first Navy memorial of American history, dedicated to the First Barbary War (1801–5), there is a list of names of the fallen, but again the names are either those of dead officers or of those who contributed to the erection of the monument (Piehler 1995: 23).

Not even the great war memorials of the nineteenth century, such as *Nelson's Column* in London and the *Walhalla* near Regensburg, pay tribute to the combatants killed in war, but rather to their generals and rulers.[36] In this epoch the contribution of the commoner in war had no part in commemorative practices. This way of memorializing war through the representation of 'great men' celebrated the cause without mentioning its main participants. Again, if there was any reference it was to statesmen and great military commanders like Gustavus Adolphus, Peter the Great, Nelson, Wellington, Napoleon, and his generals, but not to the individuals who had fought and died for their own countries.

Memorials and monuments are hence significant not only for what they represent, but also for what they exclude. In a way, the most interesting memorials of the time were those that were never built and were never meant to be erected. The lack of a wish to commemorate commoners in monuments and ceremonies stands out as one of the most important testimonies to the unproblematic nature of ordinary soldiers' death in war and of their lack of a social body.[37] Once soldiers had died, no one attempted to rescue them from oblivion.[38]

Even the French Revolution did not open the monument galaxy to humble combatants (Clarke 2007). Although the extension of military service to

war—here, indeed, to quote George Washington, 'the first blood was spilt in the dispute against Great Britain' (in Chernow 2010: 613)—the inscription on the monument lists the names of the American combatants killed in battle.

[35] For example, memorials dedicated to Montgomery, Washington, and Warren were built, respectively, in St Paul's Chapel, in New York (1787), in Richmond (1791), and Bunker Hill (1794). However, it must be noted that most American Revolutionary monuments were erected between the 1890s and the 1930s.

[36] On Nelson's memorialization, see Yarrington (1983).

[37] As an example of ancient memorial practices differing from those of the early modern era, on the Great Mound at Marathon a victory column was erected and dedicated to the 192 Greeks who died fighting the Persians in 490 BC. Another example of a memorial in the ancient world that paid tribute to common soliders is an altar, located at Adamclisi in Romania, bearing the names of nearly 4,000 Roman soldiers who probably died along the Danubian frontier during Domitian's reign (Turner 2013). However, it must be noted that in ancient Rome the war dead were rarely commemorated (Hope 2018). In this sense, the stone altar of Adamclisi is unique in the Roman world.

[38] A sort of battlefield tourism blossomed after the end of Napoleonic Wars. Waterloo, especially, became a site of mass tourism for those who wanted to see where Napoleon was routed (Semmel 2000). However, it must be stressed that Waterloo was not a site of pilgrimage but rather a popular tourist attraction, especially for British people.

all able-bodied citizens and the glorification of soldiers were simultaneous (Nef 1950; Morrissey 2014), such glorification did not translate into public representation and memorials. Indeed, though Benedict Anderson (1991: 9) was certainly correct in maintaining that 'no more arresting emblems of the modern culture of nationalism exist than cenotaphs and tombs of Unknown Soldiers', the ideology of nationalism, originally articulated in the French Revolution and powerfully employed by Napoleon on the European battlefields, did not translate immediately into commemorative devices meant to honour those who had perished for the nation. While the French soldier was portrayed as a national hero, the rhetoric was not translated into monuments. Although French citizens were expended in larger and larger numbers, their sacrifice was not recorded on war memorials. No common soldiers' names are to be found on the most important monuments dedicated to Napoleon's victories, either in the *Arc de Triomphe*, whose inscriptions refer only to Napoleon's generals and battles (Troyansky 1986b: 351), or in the other monuments dedicated to the *Grande Armée*.[39]

This point must be emphasized, because it shows that the political transition from the French absolute monarchy, based on the idea of the divine right of kings where people were subjects and soldiers were mere property of the armies, to the mass-based regime born with revolutionary France, ideologically and politically based on the will of the nation, did not change attitudes towards soldier deaths. While the French Revolution and Napoleonic campaigns changed many aspects of war in the old continent (for example, the size of armies, their organization, the political commitment of troops, and a more equal system of compensations for veterans[40]) and reintroduced the concept of the citizen-soldier in Europe, as regards the subject matter of this research it should be concluded that they did not contribute any radical transformations.

Intermezzo: The End of an Epoch and the Birth of a New One

The historical period covered in this chapter is a long age that witnessed many important developments impacting on war in the fields of technology, tactics, strategy, politics, and economics, as well as a fundamental change from wars

[39] The only noteworthy element of commemoration in the Napoleonic armies was the custom of including the name of the killed in the roll call.

[40] Although never fully implemented for fiscal and administrative reasons, the law of 6 June 1793 introduced a more equal and widespread system of benefits for veterans and pensions for war widows (Woloch 1979: 78–101).

triggered by personal disputes 'between individual princes over rights of inheritance' to conflicts between nation states (Howard 1976: 21). For the sake of the present study, however, it appears as a single epoch in which attitudes towards soldier deaths remained largely unchanged. The figure of the soldier in that period went through a major shift—the freebooting mercenary of the sixteenth and seventeenth centuries would gradually turn into the professional and 'uniformed automaton of the eighteenth century' (Showalter and Astore 2007: p. xi)—yet the meaning attached to soldier deaths remained fundamentally unaltered: the death of common soldiers was neither lamented nor celebrated. Because armies were considered anonymous collections of men rather than formed of single individuals, ordinary soldiers could be massacred on the battlefield over and over again down the centuries without raising any moral, political, or social reaction. They were passive instruments of statecraft that could be replaced by other human flesh. Their welfare was of little concern to their commanders as long as it did not diminish the strength of the army. At this time the death of a soldier was regarded as a purely physical event, without any legal, political, or social effects: it was nothing but *bare death*.

This is also, however, an epoch in which significant changes concerning attitudes towards death took place outside the area of war, in society at large. During the Middle Ages and in the early modern period death was an integral part of the daily experience of most human beings, who largely lacked the affective reactions typical of the current era. Doubtless people feared their death and attached importance to it, but, tellingly, graves were located among the living. The dead were mainly buried in churches (the clergy and local landed families), in churchyards (the poor), or in their proximity (the unbaptized and suicides), places that were sited, in turn, at the centre of communities rather than outside the city walls (Rugg 2000: 265). This burial practice has been described as 'the old regime' of the dead (Laqueur 2015: 112–210) and as the 'Christian parenthesis' (Galinié 1996: 17) between a past and a future in which the dead were physically separated from the living.[41]

By the end of the eighteenth century, the intimate conviviality with the dead ended with a change in the location and structure of burial sites.[42] Graves were removed from churchyards and within village and city centres and located on the outskirts and in the countryside, where there was space for individualized

[41] It must be noted, however, that with the progressive urbanization of American and European cities the modern cemeteries would soon be reincluded in the urban environment.

[42] In his study on death in Paris, Pierre Chaunu (1978) suggested that in the 1740s there had already set in a decline in the habit of burying the dead in churches, which therefore predates the royal decree that prohibited inhumation in churches (1776) and the closure of the cemetery *des Innocents* in 1780.

burial (Etlin 1984; Malone 2017). The movement from intramural to extramural cemeteries was accompanied by what might appear to be two contradictory features, but which are in fact intimately connected. On the one hand, the dead body became an object of fear and horror, which needed to be 'segregated' and separated from the living: after centuries of cohabitation with the living, corpses were now seen with distaste and repulsion. On the other hand, a real cult of the dead and of individual tombs developed (Richardson 1989). Building cemeteries outside the city walls in a country setting was regarded as the most decent way to dispose individually of the corpses of the dead. Indeed, the function of the modern cemetery goes beyond the mere necessity to house the deceased. On this point, Ariès (2009: 74) maintains that 'the piety and the new respect shown for tombs resulted in an extension of the surface area of cemeteries, because it had become intolerable and forbidden to pile up corpses as in the charnel houses of the Middle Ages'.

The individual tomb became the rule rather than the exception, but it was not religious devotion that lay at the basis of 'the one-grave-one body approach' to internment, but rather the need to provide a decent place for visits by those who were left behind (Wood and Williamson 2003: 18).[43] People

> wanted to go to the very spot where the body had been placed, and they wanted this place to belong totally to the deceased and to his family … People went to visit the tomb of a dear one as one would go to a relative's home … Memory conferred upon the dead a sort of immortality which was initially foreign to Christianity.
>
> (Ariès 2009: 72)

While before the eighteenth century even for the higher-ranking dead funerary memorials did not mark their buried bodies (Tarlow 1999: 124) and there is no indication that the graves of the loved ones were visited by the survivors (Ariès 1981: 475–556), in the nineteenth century tombs and cemeteries became a true obsession for the living, though obviously only for those who could afford a tomb and a decent funeral (Laqueur 2015: 312–36).

This new attitude was based on the dramatization and romanticization of the death of others, especially of family members and friends, 'an attitude which valued the sentiments of the survivors' (Wells 2000: 64–5) and which included

[43] It is significant that in his address for the dedication of Mount Auburn in September 1831, Justice Joseph Story described the contemporary Christian approach to burial as unfortunate and praised the Egyptian, Greek, and Hebrew practices (French 1974: 46).

the idea that in the corpse remains part of the identity of the deceased, something that should be preserved in death. Such a change privileged the viewpoint of the bereaved rather than that of the dead. Being the repository for the identity of the departed, corpses needed a decent and individualized burial with specific rituals for their disposal, and not the mass graves used in the past. Cemeteries were no longer merely places of burial, but sites for grief where the bereaved 'needed to be protected from inappropriate activity' (Rugg 2000: 264). Not only should the departed lie in peace, but the survivors too should be left undisturbed when visiting their loved ones.

The result was the development of a cult of cemeteries unknown in the Middle Ages or in the early modern period.[44] The individual tomb was meant to replace the deceased and restore his or her personal identity denied by death. The elaborate and eclectic shape of grave markers, the construction of funeral monuments, the gradual increase in the number of epitaphs, obituaries, and biographical sermons during funerals, and the custom of leaving floral tributes on graves, were all meant to confer a lasting and individual memory of the dead. Symbols of mourning, like the weeping willow, gradually replaced symbols of damnation and resurrection such as the skull, the flame, and the hourglass. The emphasis was on the individual, and the key relationship was no longer vertical, between the departed and God, but horizontal, between the dead and the living (Hijiya 1983: 354). Or, as Sarah Tarlow (1999: 127) put it, it was based on a 'synchronic relationship' between the living and the dead: the focus was on the survivors' grief rather than on the soul and the destiny of the dead. It is during this period that we witness another change in funerary monuments and tombs: social status in life gets replaced by familial relationships.[45]

By distancing the cemetery from towns and cities, the boundaries between the world of the living and the world of the dead were redefined. But, as noted, distancing did not mean lack of interest in the dead. Quite the contrary, the new cemetery provided a decent, beautiful, sweet-smelling place of burial, lessening the affliction of the mourner. Only by building cemeteries in such a way could the anguish and terror of death be softened. Accordingly, location and accessibility became very important issues in the planning and construction of modern cemeteries. Tellingly, when in the mid-1860s Baron

[44] On burial practices in the Middle Ages, see Lauwers (2005).

[45] In his monumental work on the family in England, Lawrence Stone (1977: ch. 6, 'The Growth of Affective Individualism') defined this new familial attachment as 'affective individualism'. Interestingly, although a harsh critic of Ariès, Stone (1981: 252) agreed with the French historian's characterization of attitudes towards death in the eighteenth and nineteenth centuries.

Georges Haussmann proposed the construction of a necropolis in Méry-sur-Oise, Parisians protested and blocked the project because the location was too far from the city (Troyansky 1986a: 173; Burton 2001: 134).

A new concept of the cemetery was introduced with the establishment of easily accessible garden and rural cemeteries in the US and in Europe, and then with the construction of memorial parks in whose design and even words the morbid connotations of the term 'cemetery' were utterly eliminated (Sloane 1991).[46] Both types of cemetery were meant to conceal death through the creation of a pleasing and peaceful landscape for the remembrance of the loved ones as opposed to the overpopulated town graveyards of the past.

Not only did this cult of the dead arise almost simultaneously throughout the western world accommodating national differences, but it gradually permeated both the middle class and the poor, who in the past, as mentioned, had generally been excluded from decent funerary and disposal practices. By the end of the nineteenth century, the middle class adopted modes of mourning and embraced funerary rites that had previously been restricted to the prosperous part of society (Llewellyn 1991: 91). While for most of the nineteenth century only the well-to-do could afford an individual grave, in the second half of the nineteenth century, burial insurance in England gradually became popular and was used to purchase grave plots and, in so doing, ensure ownership of the body of the deceased and access to full mourning and commemorative rites. It is also significant that poor families were ready to spend a considerable portion of their small wealth on a grave, which was often the only 'piece of land' paupers would ever control in their life (Strange 2003: 147).[47]

As I will argue in more detail in Chapter 3, this shift in attitudes towards death at the societal level not only underlay the individualization of burials and the origins of modern funerary practices, but seems also to be at the root

[46] The first rural cemetery in the US was the Grove Street Cemetery in New Haven, established in 1796. Père Lachaise in Paris was the first garden cemetery in Europe, opened in 1804, followed by the Necropolis in Glasgow (1832), Kensal Green (1832), and Highgate (1839), and Abney Park in London (1840). The most famous American rural cemetery is Mount Auburn in Cambridge, Massachusetts, which, after its dedication in September 1831, became the model for garden cemeteries around the western world. In 1817, 1851, and 1866, respectively, monumental cemeteries such as la Villetta in Parma, Staglieno in Genoa, and the cemetery of Milan were established, the last-named representing in Italy's case 'the most important example of … the confinement of death within the artistically modelled boundaries of the extra-urban cemetery' (Canella 2010: 129). Although these Italian cemeteries were built as monumental structures rather than landscaped gardens (Malone 2017), they did not differ from the latter in building a new relationship between the dead and the living.

[47] It has been argued that the widespread desire for a respectable death was the result of 'a materialistic expression of success' (Mytum 1989: 295). However, there is no rational reason why burial and the funeral were considered important areas of status expression. That people risked financial ruin and compromised other areas of life for what was considered an appropriate burial and funeral should make us question a purely materialist explanation.

of military cemeteries in the second half of the nineteenth century and the inclusion of ordinary soldiers in war commemoration. In short, the change in attitudes towards soldier deaths that I will describe in the next chapter seems to mirror the development of attitudes in society.[48]

[48] As we shall see, it was no coincidence that many Civil War cemeteries took on the characteristics of civilian cemeteries (Piehler 1995: 50). In Germany, France, and Italy too, First World War cemeteries were often designed as garden cemeteries. The *jardins funèbres* in France, the *Heldenhaine* (heroes' gardens) in Germany, and the *giardini delle rimembranze* (remembrance gardens) in Italy, where plaques with the names of the war dead were attached to trees, all showed how societal attitudes towards death permeated the issue of burying fallen soldiers.

3
Sacrificial Death

For the purpose of this investigation, the period that separates the end of the Napoleonic Wars from the Second World War marks a transition embodying contradictory trends. On the one hand, combat on the battlefield was still based on indifference to the sufferings and lives of common soldiers, who were treated like an anonymous mass that could be sacrificed according to the requirements of the war theatre in question. On the other hand, three important changes regarding both the natural and the social bodies of soldiers materialized: the wounded received better treatment than in the past, there was a systematic attempt to dispose of the bodies of the war dead in decent, individualized form, and ordinary soldiers were spotlit in war commemoration.

The origin of both modern military cemeteries and the systematic inclusion of ordinary soldiers in commemorative practices can actually be traced back to the second half of the nineteenth century. Not only—unlike the early modern period—did states begin to have cemeteries built for their war dead, but when feasible they also provided the fallen with individual tombs. Such a marked change in the disposal of the war dead, which took place in both the United States and Europe, had a threefold nature: soldiers' corpses ceased to be considered as mere flesh that could be either left to rot in the open air or interred in mass graves; ideally, bodies were supposed to be buried in an individualized way; and military cemeteries were endowed with national and political meaning. At long last, the average soldier was included in public rituals of commemoration. As we shall see, it is a paradox of late-nineteenth-century warfare and the First World War that soldiers should still be treated as inferiors on the battlefield, while they were glorified and even sanctified in death and memory.

In order to show both continuity and change in how soldiers were employed on the battleground, disposed of, and commemorated once dead, some of the most important conflicts of the nineteenth century will be outlined, especially the Crimean War, the American Civil War, and the Franco-Prussian War.[1] In

[1] On the importance of these three armed conflicts for nineteenth-century history, see Holden Reid (2006).

Life, Death, and the Western Way of War. Lorenzo Zambernardi, Oxford University Press.
 DOI: 10.1093/oso/9780192858245.003.0004

relation to the first half of the twentieth century, the research will focus on the First World War. As the message to be conveyed about the decades at issue has to do with gradual changes and not unbroken continuity, the narrative concerning the three types of evidence is still organized thematically within this chapter, but it is also given as a chronology to show how the meaning attached to soldier deaths has evolved historically.

The Natural and the Social Bodies of Soldiers in Mid-Nineteenth Century Warfare

In the nineteenth century, soldiering was still largely an activity for men who had little to offer to society. Although the French Revolution contributed to opening the officer corps to commoners, after the end of the Napoleonic Wars most European armies restored the traditional practice of recruiting officers from the aristocracy and ordinary soldiers from the humblest part of society (French 2005: 75). Most men enlisted for the usual motivations: necessity, inability to find a better job, bribery, coercion, and the like. On the British soldier, historian Albert Tucker declared that he was 'unskilled; and like the unemployed of the workhouse, he should ... be offered only the minimum of pay and the barest of necessities in living conditions' (in Neiberg 2006: 106). To boot, serving in the rank and file still implied a social stigma difficult to escape.

In European countries such as Austria–Hungary, France, and Russia, whose armies were based on conscription, members of the wealthy class could buy substitutes. Although the military in the United States were on average better educated than in Europe, the American soldier in the mid-nineteenth century was almost as low born as his continental counterpart. Because in time of peace Americans were still enlisting from sheer necessity (Shy 1976: 30–2), the army was formed mostly from the least fortunate members of society. In time of war, militiamen and volunteers were added to regulars, but human qualities were not greatly improved. In a telling account, Daniel Harvey Hill, a regular artillery officer during the Mexican War, described Louisiana Volunteers as 'a lawless drunken rabble' (in Herrera 2001: 38). Not surprisingly, in the several minor and the few major wars of the nineteenth century, European and American statesmen and generals alike continued to care little for their men as individual human beings. The protection of soldiers' lives on the battlefield was not a primary concern.

The Crimean War (1854–6) is a striking example of the indifference to soldiers' lives that characterized European political and military leadership in the mid-nineteenth century. The human cost of the conflict was enormous: around three-quarters of a million soldiers died during the war (Figes 2010). French casualties amounted to approximately 95,000, about 30 per cent of the troops dispatched to Crimea; Russia lost around 450,000 men, largely as a result of disease and malnutrition (Clodfelter 2008: 194); British fatalities were appalling too: 20 per cent of the men sent to the Crimea died there (Gill and Gill 2005: 1799). Although Italy lost only 28 soldiers in combat, mainly in the battle of the Chernaya, 2,138 men died of disease (Clodfelter 2008: 194), about 15 per cent of the armed forces sent to Crimea by Count Cavour.

Besides being badly conducted by Britain—the famous Charge of the Light Brigade at Balaklava will remain a typical example of the pointless waste of men—both combat and attendance to the wounded displayed the usual lack of concern for human lives.[2] The vast majority of casualties in this war for all belligerents, as in previous armed conflicts, resulted not from fighting but rather from disease and sickness. In the Crimea, especially during the siege of Sebastopol, the abject sanitary conditions of the British army created a situation in which soldiers died four times more often from disease and illness contracted during the war than from battle wounds (Gabriel 2013: 148).

Apart from the fact that medical science had not much to offer in terms of treatment, the squalid conditions of hospitals and barracks favoured the spread of infectious diseases such as typhus, typhoid fever, tuberculosis, and cholera, which decimated the troops. Moreover, British soldiers had to take care of themselves even for food supplies. Edward Hamley (1824–93), who fought in the Crimean campaign and was to become one of the major military critics of his time, provided an apt description of the conditions that the British soldier endured: 'Coming from the trenches the men had to go far afield to seek for roots … to cook their food; it is hardly surprising that many … ate their salt pork uncooked; and … under such diet and exposure, the numbers of the sick increased' (in Ropp 2000: 167). As a result of poor diet, inadequate shelter, and bad field sanitation, disease wiped out large numbers of men in all armies. Dysentery caused nearly 50 per cent of deaths.

Curing the sick and wounded was not the main objective of British hospitals during the Crimean War; rather hospitals were primarily meant to isolate soldiers with fever from those in good physical shape: 'Soldiers were not sent

[2] Moreover, the disabled veterans of the Light Brigade were denied any long-term care, apart from the promise of obtaining beggar's licences (Gabriel 2013: 63).

to Scutari to be healed so much as to die' (Gill and Gill 2005: 1800). In 1862, the heroic nurse Florence Nightingale (2011: 342), who tirelessly tended the British sick and wounded in those hospitals, described their predicament in these words: 'war has been conducted in more or less forgetfulness, sometimes in total oblivion, of the fact that the soldier is a mortal man, subject to all the ills following on wet and cold, want of shelter, bad food, excessive fatigue, bad water, intemperate habits, and foul air.'

Although the sanitary conditions exposed by Nightingale spoke for several other previous European wars, during the Crimean conflict the flow of information from the battlefield to the home front, combined with the emergence of novel social sensitivity towards soldiers, diminished the usual disregard for the troops. Previous military accounts had tended to be published long after battles had taken place. This time the invention of the electrical telegraph by Samuel Morse and other inventors gave public opinion unprecedented access to reports from the theatre of war. In disclosing the 'dreadful conditions under which the army was suffering' (Murray 2005: 222), British correspondents blamed the military medical corps for negligence in the treatment of their soldiers and wounded. In particular, William Howard Russell's reports for *The Times* of London denouncing the plight of British wounded men who received little or no medical treatment convinced Sidney Herbert, the Secretary at War, to ask his friend Florence Nightingale to go to the Crimea with a team of nurses.[3]

Although battles in the war, like the Charge of the Light Brigade, were portrayed as beautiful and glorious acts of sacrifice for the country, as in Alfred Tennyson's poems and William Simpson's paintings, in the course and in the aftermath of the conflict a new form of heroism was emerging in Europe. As Stefanie Markovits (2012) rightly noted, the two most famous heroes of the conflict are neither generals nor combatants, but a reporter and a nurse: 'a man who wielded a pen, the other a lady who carried a lamp: not a sword nor even a musket in sight.'[4]

Despite the rise of a novel attitude to the wounded during battle and the attempt to provide a decent burial for fallen soldiers, with the construction of military cemeteries as at Cathcart's Hill and in the village of Jeni-Koi, most

[3] In the well-known photographs of the Crimean War by Roger Fenton, dead bodies are not displayed. Not only was his assignment in the Crimea sponsored by Prince Albert with the goal of providing a positive description of the war, but the War Office explicitly instructed Fenton not to portray the dead. As Sontag (2003: 50) explained, his pictures 'are tableaux of military life behind the front lines; the war—movement, disorder, drama—stays off-camera.'

[4] Not only did Nightingale and her nurses save many human lives by cleaning the floors and changing sheets and blankets; they also wrote and helped write letters to the families of the wounded, the dying, and the dead. In so doing, they performed a service that would become common practice in wartime.

corpses of war dead were still hastily tumbled into common graves located in a variety of burial sites, which vanished in the following decades. Indeed, graves individually marked represented only a tiny minority: '51 out of 324 in one of the burial grounds of the Second Brigade, Light Division; 22 out of 1,334 in the cemeteries of the Third Division' (Laqueur 1994: 151–2). In the only Crimean war cemetery that still exists, the Haidar Pasha cemetery, the corpses of the six thousand men who perished from a cholera epidemic and were buried here were virtually all interred anonymously (Summers 2010: 10).

Nevertheless, it must be emphasized that public concern for the conditions of troops marked a significant change in societal attitudes towards soldiers. The creation of the Red Cross, whose origin and main developments are outlined in the following section, is further evidence for this newly emerging sensitivity.

Changing Attitudes towards the Wounded

While most philosophical and legal speculation on *jus in bello* in early modern Europe concerned the treatment of prisoners of war and the protection of civilians—a principle that, in spite of its limited success, had been recognized at least since the Middle Ages (Bellamy 2006: ch. 2, 'The Middle Ages)[5]—a different emphasis began to develop in the second half of the nineteenth century: the soldier, especially the wounded one, became the very centre of legal codification and protection.

A small, certainly insignificant change in the short term, but with profound long-term consequences on soldiers' lives, occurred a few years after the end of the Crimean War, on 24 June 1859, in the immediate aftermath of the battle of Solferino, near Mantua, at which the French and Sardinian armies clashed against the Austrians. The battle was not only an important step in the process of Italian unification, which took place in 1861, but was also the worst military carnage on the Continent since Waterloo: approximately six thousand soldiers died in nine hours' fighting and many more were wounded. The sheer number of casualties in the field caught the medical services of both sides largely unprepared. Not only did the French have more veterinary surgeons than physicians, but it took them six days to remove more than ten thousand injured men from the field. Wandering in search of food and water, about nine thousand soldiers made it to Castiglione delle Stiviere, where Henry Dunant, at that time a Swiss entrepreneur, arrived in order to meet Napoleon III for business and financial

[5] See Kinsella (2011: ch. 2, 'Martial Piety') for a critical analysis of the principle of distinction in the Middle Ages, which stresses the differentiation of immunity and the ambiguities in their formulation and implementation.

reasons. With no medical expertise, but deeply touched by the mass carnage he saw, Dunant helped, cleaned, and dressed the wounded sheltering in the Chiesa Maggiore (Bugnion 2009: 3).

In 1862, after his return to Geneva, Dunant wrote and published *Un souvenir de Solferino*, a passionate, vibrant account recording his memories of the wounded and deploring that they had been left to themselves after being injured in action. In the ensuing years, with the help of the Swiss lawyer Gustave Moynier and other activists, Dunant pressed for the establishment of volunteer societies for the relief of the wounded in time of war. In 1863, the International Committee for Relief to the Wounded (ICRW) was created.

As the change in attitudes towards soldiers underlying the creation of the ICRW gradually grew in strength, new prospects opened up for the care of the wounded. During the Second Schleswig War in 1864, the Prussian Society provided assistance for injured combatants, and that year an international conference held in Geneva approved the first Geneva Convention for the Amelioration of the Condition of the Wounded in Armies in the Field (that is, the First Geneva Convention). The original signatory states were twelve in number: Baden, Belgium, Denmark, France, Hesse, Italy, the Netherlands, Portugal, Prussia, Spain, Switzerland, and Württemberg. In the following years, Britain (1865) and the United States (1882) joined the Convention as well.

Much of the focus of the Convention was on protecting the wounded and respecting the neutrality of medical personnel, which was vital for the care of the injured. In particular, a major legal revolution is to be found in Art. 5 stipulating that civilians who helped the wounded would be considered neutral and their liberty would be preserved. By considering the wounded as non-combatants and by declaring the neutrality of ambulances, hospitals, and their staff, the Convention laid the foundations for effective care of the wounded.[6]

By the century's end, this new attitude to soldiers, combined with the creation of sanitary corps all over Europe and medical innovations such as Joseph Lister's antiseptic surgery developed in the 1860s and 1870s, contributed to a reduction in mortality rates among the sick and wounded. Moreover, discoveries in the area of bacteriology led to better practices in immunology and in the prevention of communicable disease. Obviously, that was the case only in those armies that had introduced new medical practices, like

[6] It is worth noting, however, that a concern for the wounded and prisoners was also shown during the War of the Austrian Succession when, after the battle of Dettingen (27 June 1743), the opposing sides in war signed an agreement stipulating that medical personnel would be considered as non-combatant (Gabriel 2013: 98).

Prussia after the war against Austria, but unlike France, which forbade antiseptic surgery (Neiberg 2006: 87–8) and failed to introduce systematic inoculation for its troops (Swain 1970: 514).[7] The success of the German medical corps in reducing the death rates showed the other nations the potential of combining scientific knowledge, hygiene, and efficient administration.

Changing Attitudes towards the Wounded in the United States

This novel interest and care for the wounded was far from being the product of personal feelings on the part of Henry Dunant and his associates. Similar developments were taking place in the same period across the Atlantic, specifically in the context of the American Civil War. Two main changes concerning soldiers came to the forefront during and after this conflict. First, a variety of organizations and associations were established for the provision of medical care to the injured. Secondly, there was an attempt to introduce legal restraints to the horrors of war for combatants.

On the first point, it should be noted that, at the beginning of the conflict, both Union and Confederates were largely unprepared for the evacuation and treatment of the wounded (Greenwood and Berry 2005: 17). No ambulance system and no systematic way of tending the injured were in place. For instance, after the First Battle of Bull Run (21 July 1861) the wounded were moved on vehicles commandeered from Washington (Bollet 2002: 3). And at the Battle of Wilson's Creek (10 August 1861), owing to a lack of ambulances, the wounded were not moved for six days (Gabriel 2013: 169). However, in the following months, medical aid gradually improved with the introduction of a specialized ambulance corps and the pavilion hospital. Other organizations, such as the Christian Commission, provided floating hospitals, doctors, and nurses during and after battles, and helped families to locate their wounded relatives in the Union army.

In this connection, the work of Dorothea Dix is worth mentioning briefly. An activist for the care and humane treatment of the mentally ill and disabled in the pre-Civil War period, she was appointed by the Union as Superintendent of Army Nurses (June 1861) and assigned the task of arranging assistance for the sick and wounded of the Union. Another prominent figure in the tending of the wounded was Clara Barton, also known as the 'angel of the battlefield'

[7] According to Keegan and Holmes (1985: 150), the Boer War (1899–1902) was 'the first conflict in which a soldier who underwent an amputation was more likely to survive the operation and its effects than to die of gangrene'.

(Egenes and Vlasses 2016: 71), who not only greatly contributed to treatment of the injured, but after the Civil War would run the Missing Soldiers Office and was to become the first president of the American branch of the Red Cross. Moreover, after the war had ended, a comprehensive pension system was created, which greatly helped not only Union veterans and the disabled, but also orphans, widows, and other dependants (Rostker 2013: 102–6).

As regards warfare, on 24 April 1863 President Lincoln promulgated General Orders No. 100 (*Instructions for the Government of Armies of the United States in the Field*) with the purpose of regulating the military conduct of the Union armies. The *Instructions* is known as the Lieber Code, after the name of the German-born Columbia College professor of law who authored it, Francis Lieber. The rules of the Lieber Code were meant to govern the conduct of troops in relation to both enemy combatants and civilians. Contrary to some long-standing views, the Lieber Code did not make wanton violence against civilians illegal. The Code prohibited only violence against civilians and their property when 'not commanded by the authorized officer' (Art. 44). One of the most significant points of the Code concerned the treatment of wounded enemies and prisoners of war. In particular, Art. 61 explicitly stipulated that 'troops that give no quarter have no right to kill enemies already disabled on the ground, or prisoners captured by other troops'. Although the principle providing for the protection of prisoners of war and the wounded was far from being new,[8] its legal formalization in the Code counts as further evidence that attitudes towards soldiers' lives were changing. In fact, the document turned an existing moral convention into a legal military regulation.

It is difficult to establish whether the Lieber Code actually restrained warfare during the Civil War. While it was certainly easier to formulate laws of war than to enforce them, especially in the territory under Confederate authority,[9] the Code was quite correctly described as 'the first set of war rules to show genuine concern for the enemy' (Robertson 2002: 183), and it also acted as a legal reference for accusing and sentencing war criminals after the conflict.[10]

Whatever the practical merits and limitations of the Lieber Code, it is significant for our enquiry that protection of the wounded was included in

[8] On the treatment of prisoners of war in early modern Europe, see, e.g., Alberico Gentili, *De iuri belli libri tres* (2008), originally published in 1589, notably bk II, chs 15–18.

[9] Confederates did not grant the status and rights of prisoners of war to African American soldiers, who could be massacred even when they had surrendered, as happened at Fort Pillow in April 1864.

[10] After the Civil War, a number of Southern officers were tried and executed for war crimes on the basis of the Lieber Code, including Henry Wirz, commander of the infamous Confederate prison Camp Sumter near Andersonville, where 13,000 inmates died of disease and violence. Overall approximately 30,000 Union prisoners and 26,000 Confederates perished in Civil War prisons (Mitchell 1997: 571, 577, 583).

legal codification both among states (First Geneva Convention) and within states (the Lieber Code itself), since it indicates that the next step towards regulation of warfare was again meant to diminish the sufferings of soldiers. Thus, the use of explosive bullets was banned in 1868 by the St Petersburg Declaration, because, as explicitly stated in the preamble, 'the only legitimate object which States should endeavor to accomplish during war is to weaken the military forces of the enemy' (in Hayashi 2020: 58). Killing in war was still accepted as a perfectly legitimate practice, but it was to come about in such a way as to lessen the pain of combatants.

In spite of the improvements in the way the wounded were tended on and off the battlefield, no significant change took place in the degree of soldier expendability in battle. The American Civil War, the wars for German unification, and the First World War were to show once again how little western governments were concerned about the death of their combatants.

No Change to the Natural Bodies of Soldiers

The American Civil War

The Civil War is often seen as a historical preview to the atrocities that were committed in the mass slaughters of the Russo-Japanese War (1904–5) and the First World War (Hagerman 1988). It indeed anticipated some of the main features of twentieth-century combat: earthen field fortifications, mass attacks, the killing zone, and a combination of popular mobilization and mechanization. Moreover, innovations like the conoidal bullet, used for the first time in the Crimean War, greatly improved the accuracy and lethality of rifled muskets, which could now kill even at 1,000 yards (Dupuy 1984: 191). In this respect, the distinguished military historian Williamson Murray (2005: 223) maintained that the American Civil War 'ranks as the most important conflict of the nineteenth century because, for the first time, opposing governments harnessed the popular enthusiasm of the French Revolution to the industrial technology that was sweeping the West'. Undoubtedly, such a combination turned the Civil War into a horrific, deadly conflict.

The war was long, ferocious, and exhausting. Apart from the Taiping Rebellion (1850–64) in which about thirty million people were killed, the Civil War was the bloodiest armed conflict of the nineteenth century (Degler 1997: 61). Approximately 620,000 soldiers died from 1861 to 1865, as many as all the American fallen in the Revolution, the War of 1812, the Mexican War, the Spanish–American War, the First and Second World Wars, and the Korean

War (Murray 1996, 15). As in previous conflicts, the majority of men died from disease and exposure, approximately 225,000 for the Union and 90,000 for the Confederates (Congressional Research Services 2019: 2–3). Hence, the mortality rate in the Civil War was much higher than in other major conflicts involving the United States. The overall figure was six times as high as in the Second World War. 'A similar rate, about 2 percent, in the United States', writes historian Drew Gilpin Faust (2008: p. xi), 'would mean six million fatalities' today.

The difference between the South and the North in terms of industry, infrastructure (especially rail transport), and men meant that the Confederates were doomed to succumb in a prolonged war. Besides having a population of twenty-two million in comparison to the less than six million white residents in the South, the North possessed nearly all the industrial resources of the country.[11] That is why the outcome of a protracted conflict was almost inevitable.

From the start the Union tried to strangle the South by imposing a blockade on Confederate ports and by cutting the South into two different areas divided by the Mississippi river. Although sea power played an important role in the overall development of the war (Dupuy and Dupuy 1956: 292–4), the plan did not bring the South to its knees. Costly battles characterized by massive assaults, as at Antietam (17 September 1862), proved crucial in stopping Confederate General Robert E. Lee's Army of Northern Virginia from advancing into the territory of the Union.

The war was fought in a variety of ways depending on the theatre of operations and the development of the conflict itself, but, as a general trend, it grew costlier and more devastating as it progressed, so much so that five of the six bloodiest battles took place after April 1863. Not only was battle unavoidable, but inflicting major casualties on the opponent was seen by both sides as the best strategy to end the war quickly. In pursuit of a decisive victory, General Lee tried brutal frontal attacks in the three-day battle of Gettysburg on 1–3 July 1963. The outcome was horrific: seven thousand men were killed in action, and the fatally wounded who perished on the following days, combined with the missing, greatly increased the number of casualties (Clodfelter 2008: 300–1). As 1864 was an electoral year, Lee hoped that a large number of dead and wounded men in the ranks of the Union would convince the Northern population not to re-elect Lincoln.

[11] The 1860 United States Census determined the American population to be 31,443,321, including almost four million slaves. However, on the limits of population statistics to measure potential manpower during the Civil War, see Dupuy and Dupuy (1956: 215) and Weigley (1973: 129–30).

The strategy failed because the North was willing to accept great sacrifices and sustain heavy casualties to achieve victory. General Ulysses S. Grant was not shy of expending his men and, having a larger number of soldiers than Lee, was not greatly concerned about taking losses. Inflicting heavy casualties on the Confederates was one of the most effective means of ending the war, and suffering heavy casualties was the price to pay in order to prevail.[12] As a result, the proportion of casualties increased to 21.3 per cent for the Union (Dupuy 1984: 171), exceeding those of Napoleonic warfare.

It was Grant's military goal to inflict cumulative damage on the enemy through a campaign of attrition. He was not, as some historians argued, especially indifferent to his men's lives; he was simply a general of his time, ready to sustain horrendous casualties in order to achieve military victory. Because succeeding in the bloody trade of mid-nineteenth-century warfare meant accepting mass death, even in the face of huge losses both Grant and Lincoln never lost their iron resolution in expending immense numbers of men. Unlike the South, the Union won the Civil War because it could sustain the economic and human costs of a long and bloody conflict of attrition.

That does not equate with suggesting that both sides in the conflict unreasonably sacrificed their men. As usual, numbers meant military strength. Hence, as in prior wars, soldiers were spread out on the battlefield and ordered to take advantage of cover whenever possible—for innovations in weaponry in the Civil War made cover from enemy fire even more important than before. The result of the 'greater deadliness, range, and accuracy of weapons' was to drive 'men to earth' (Dupuy and Dupuy 1956: 219), especially in the final phase of the conflict when entrenchments changed the face of battle (McWhiney and Jamieson 1982: 73–5).

However, because a few entrenched men could hold in check a more numerous enemy, the attacking forces had to resort to mass assaults in order to prevail. Not surprisingly, 'the frontal attack against an entrenched, determined, foe became more and more suicidal' (Dupuy and Dupuy 1956: 219). Consequently, casualty rates in battle were high, especially for the Union. Grant's armies had an average casualty 'of 3:2 vis-à-vis those of Lee's forces' (Sondhaus 2006: 55), but the death ratio could be higher. In a series of battles that took place in the summer of 1864 just before the siege of Petersburg, the Union lost 65,000 soldiers, dead or wounded, which almost totalled the size of Lee's army (Clodfelter 2008: 310). In one of the clashes, at Cold Harbor,

[12] Drawing upon Hans Delbrück's famous distinction between two kinds of military strategy (see Craig 1986: 341–4), historian Russell F. Weigley (1973: 128) described Grant's approach as a 'strategy of annihilation'.

the Union suffered 13,000 casualties against 3,000 among the Confederates, and 7,000 Union soldiers were killed in only thirty minutes (Hess 1997: 486). Such a large number of Union casualties resulted from suicidal infantry assaults against Southern trench fortifications. In other words, if the exigencies of war required it, Union forces were every bit as expendable as their European counterparts.

The Wars of German Unification

In the second half of the nineteenth century, changes in logistics and military technology significantly affected warfare on the European continent. 'Railroads and rifles', to quote the title of an important study on the wars of German Unification (Showalter 1975), decided the major European conflicts in the second half of the nineteenth century. On the one hand, the breech-loading needle gun used by the Prussians proved to be an important advantage in 1866, since it could be reloaded while a soldier was prone in defence (Degler 1997: 67), unlike the Lorenz rifles used by the Austrians, which had to be reloaded from a standing position. On the other hand, railroads made it possible to mobilize unprecedented numbers of men, who could be moved quickly and en masse to the battlefront. During the Franco-Prussian War of 1870–1, Helmuth von Moltke, the chief of the Prussian General Staff, succeeded in bringing together 1.2 million men, double the size of the army that Napoleon had deployed to invade Russia (Howard 1976: 99).

Along with logistics and military technology, field tactics changed as well. With more and more deadly weapons, close formation had temporarily to be replaced by dispersion. During the Franco-Prussian War, in order to reduce the target for French firepower, the Prussians employed their men in disperse formations that could outmanoeuvre the thicker and slower French columns and lines (Black 2005: 64). Moltke tried to replace frontal attacks with a strategy of enveloping the enemy forces, which was designed to compel the adversary to resort to costly onslaughts. However, bloodbath could not be avoided, and soldier expendability remained high in that conflict too. For example, at 'Morsbronn, the French cavalry was slaughtered in a single charge; and on 6 August at Froeschwiller, those killed or wounded exceeded 20000 on the two sides' (Capdevila and Voldman 2006: 41). Not coincidentally, the Franco-Prussian conflict was the first war in which the number of combatants killed by enemy fire exceeded the number of those who succumbed to disease (Gabriel 2013: 133).

Brute figures apart, Michael Howard's vivid portrayal of the battle at Gravelotte-Saint-Privat near the fortress of Metz on 18 August 1870 gives an impressive glimpse of the carnage entailed in late-nineteenth-century warfare:

> the skirmishing lines of the [Prussian] Guard, with thick columns behind them, extended themselves over the bare fields below St Privat and began to make their way up the slopes in the face of the French fire … The result was a massacre. The field officers on their horses were the first casualties. The men on foot struggled forward against the *chassepot* fire as if into a hailstorm … All formations disintegrated … The casualty returns were to reveal over 8,000 officers and men killed and wounded, mostly in twenty minutes; more than a quarter of the entire corps strength.
>
> (Howard 1961: 138)

As this historical account clearly illustrates, war implied taking whatever casualties would result. The increase in firepower meant that attempts by soldiers to close with their entrenched enemies led to wave after wave of casualties among the attackers. Not coincidentally, the Franco-Prussian War brought about the extermination of more than 180,000 French and German men in a period of only six months. What is more, despite heavy losses in battles such as Saint-Privat, tactical regulations after the war once again emphasized the importance of close-order formation rather than dispersion over the zone swept by enemy fire (Holmes 2005: 232). In other words, the life of soldiers on the battlefield was still a secondary object in the great game of war.

So far, this enquiry has ascertained that in the second half of the nineteenth century, while the wounded received better treatment than in the past, combatants were still cheaply expended. Hereafter, it will be argued that in this period the real revolution concerning the meaning attached to soldiers' life and death concerned not the natural body of combatants but rather their social body, as shown by the disposal of their corpses and commemorative practices.

Military Cemeteries in the United States

In 1847, the remains of forty-two American soldiers who had fallen in the war with Mexico (1846–8) were recovered from their battlefield graves and then laid to rest in the state cemetery of Frankfort, Kentucky. Three years later, in 1851, the first American war cemetery was established in Mexico City, where the bones of officers and common soldiers were moved from the city garbage dump and buried in a mass grave (Mallett 2013: 443). Notwithstanding these significant examples, it was the American Civil War that led to a radical break in the history of the disposal of the war dead. While only 6 per cent of the fallen in the Mexican war were retrieved and reinterred, the issue of locating the dead, identifying their remains, and burying their corpses in military cemeteries was a central policy during the Civil War and in its aftermath.[13] Whereas US Quartermaster General Jessup's annual report of 1848 after the conflict against Mexico makes no mention of military burials (Steere 1948: 160), the identified remains of soldiers in the Civil War totalled 54 per cent (Faust 2008: 236).

In 1862, when warfare was intensifying and casualties were mounting, Congress empowered the president to purchase burial grounds for those perishing in combat. In the same year, the War Department requested the Quartermaster General to provide military hospitals with appropriate forms on which the war dead could be recorded. In April 1862, 'in order to secure, as far as possible, the decent interment of those who have fallen, or may fall, in battle', General Orders No. 33 established that

> commanding generals to lay off lots of ground in some suitable spot near every battle-field, so soon as it may be in their power, and to cause the remains of those killed to be interred, with headboards to the graves bearing numbers, and, where practicable, the names of the persons buried in them. A register of each burial ground will be preserved, in which will be noted the marks corresponding with the headboards.
>
> (in Sledge 2005: 33)

The goal was twofold: first, giving proper, dignified burial to the dead and, secondly, identifying their mortal remains. Since anonymous mass graves were now associated with a sense of ignominy and were considered contrary to standards of decency, a series of cemeteries was created in places such as Arlington, Gettysburg, and Antietam, where at least some of the fallen could be saved from oblivion.[14] Arlington was built by the Union, Antietam by a private association (the Maryland Unionist), and Gettysburg by Pennsylvania and other states (Piehler 1995: 49–52). Whatever the origins of military cemeteries and the actors involved in their construction, there was a shared sustained effort to bring the corpses of ordinary soldiers back from the dump of animal carcasses to the path of humanity.

Not only had commanders in the field been charged with the duty of identifying and burying the war dead, but, with the introduction of the Graves Registration Service in July 1864 (Steere 1948: 151–6), disinterring, collecting, and identifying fallen soldiers and relocating their remains had also

[13] The Mexican war was a medical disaster for the United States and in statistical terms the deadliest conflict from disease for American troops. While only 1,458 soldiers were killed in action, 10,790 died of disease (Gabriel 2013: 162).

[14] About half of the Union fallen and two-thirds of the Confederate dead at Gettysburg were interred as unknown (Coco 1995: 93).

become a governmental responsibility and a federal policy, which created a massive programme of reburial and the establishment of several national cemeteries.[15]

As only single burials could secure the identity and remembrance of the deceased, Congress established in 1866 that all Union soldiers were to be buried in a permanent individual way (Piehler 1995: 168). Thanks to that innovation, graves in Civil War cemeteries are numbered, and the individuality of the dead, when known, is preserved in written records of the names of the buried. Although Civil War soldiers had been expended as an amorphous, anonymous collectivity, recording their names marked out each and every one of the known dead, turning them from mere uniforms and numbers into persons with a right to be remembered as single individuals.

The novel societal attitude is also apparent in soldiers' fear of being interred anonymously. Not only did survivors ask to identify the corpses of their loved ones; combatants also expected their identities to be preserved after death. To quote one meaningful example: a Northern war correspondent, William Swinton, recounted that, 'when [General] Meade deployed his forces before Lee's field works paralleling the ravine of Mine Run and ordered the V Corps to deliver a frontal attack, soldiers of the assault force, well aware of the cost of such an operation, carefully examined their equipment, then wrote their names on slips of paper and pinned them to the their blouses' (Steere 1948: 156). The significance of this gesture should not be underestimated, since it manifested an ideal of remembrance and an explicit desire to save soldiers' identities after their likely death.[16] It also showed that death was increasingly being thought of as the risk of being forgotten. Although official identification of US servicemen through aluminium dog tags would not be introduced until the First World War, Civil War soldiers could either buy belt buckles and badges with their names inscribed or improvise by writing their personal details on pieces of paper stuck to their uniforms. Even if mass burials made it hard to identify the fallen, personal items, such as letters and diaries, were left on the bodies in order to make identification possible when they were disinterred (Faust 2008: 71–2).

When the conflict was over, however, a huge number of corpses remained unidentified: approximately 170,000 Union dead were either missing

[15] During the war, the Quartermaster Corps was the institution in charge of looking after Union dead. After the war, the Cemeterial Division was responsible for disinterring and reburying the fallen (Anders 1988).

[16] The fact that the attack was cancelled does not reduce the importance that soldiers themselves attached to the identification of their bodies in case of death.

or unnamed at the end of recovery operations in 1871. The sheer quantity of bodies, combined with the unpreparedness of the administrative machine, made it impossible to achieve better numbers in the identification and decent burial of soldiers. The fact that both North and South failed to keep records of the dead or to provide respectful burials for many of the fallen, proves only that there was a lack of administrative capacity rather than that there was the traditional disregard for the war dead. As Faust (2008: 65) explained, the 'structures and resources that would have been necessary to implement such policies were hardly even imagined, much less provided: the Union had no regular burial details, no graves registration units, and until 1864 no comprehensive ambulance service'. As the bodies could not be recovered immediately after battle, decomposed remains became difficult to identify and required a forensic expertise that was not available at the time.

An administration that was up to such a massive undertaking was not developed until after the conflict. Especially when the intensity of battles escalated, the system proved inadequate, and a number of private actors, such as undertakers, embalmers, and companies like Staunton Transportation and the Adams Express, made a profit from assisting the families who were trying to retrieve their loved ones for interment at home.[17] Although making money out of the dead might appear to be offensive, a real business developed in the years of the Civil War.[18]

Thus, it is not surprising at all that, despite considerable efforts to provide decent burial for the fallen, cadavers were often left to stink and blacken for weeks on the battle sites, as in earlier times. In the months following major battles, it was common for soldiers on the move to find large quantities of human bones and skeletons, either because the dead had been left on the surface of battlefields or because the burials were so shallow that rain had washed up the corpses, which was particularly unfortunate in that exposure to natural elements could make identification extremely hard, if not totally impossible.

[17] For $25–$50, embalmers prepared corpses for the journey home (Laderman 1996: 109). The new respect for soldiers' corpses was a real godsend for this new type of business, which continued to expand in the post-Civil War period and turned into a big commercial enterprise in the twentieth century. The origin of the American funeral business and of the so-called commodification of the dead, with embalming at its centre, dates back to the years of the Civil War and especially its aftermath (Farrell 1980).

[18] Embalming was meant to delay the repulsive bodily decay and present the cadaver in a lifelike state, so that the deceased could be displayed during funeral rites. From this point of view, the war was a contributory factor in changing some of the mortuary practices surrounding the corpse. Because war impaired normal funerary and disposal rites, Faust argues that embalming, which allowed bodies to be moved across wide geographical spaces with no health danger, was an attempt to restore 'good death' in war. However, 'good death' was traditionally conceived of as an accepted rupture, not a blurring of the boundary between life and death. That is why Geoffrey Gorer (1965: 196) polemically described 'the art of embalmers' as 'an art of complete denial'.

There was a further reason for the amount of human flesh and bones that could be found on the fields: in order not to lose a military advantage, truces for removing the dead were often rejected. Even in an age of great changes in attitudes towards death, military needs trumped the dignity of the fallen. As General Grant put it during the battle of Cold Harbor: 'Lee was on his knees begging for time to bury his dead. But in this cruel war the business of generals is with the living' (in Goldfield 2011: 329). Mass burials were therefore used, especially for enemies, who were often placed in trenches, usually without any covering and more rarely rolled in blankets and then laid head and feet alternately as a way of saving space.

Despite administrative shortcomings in the disposal of bodies, the improvements in the treatment of soldiers' corpses are striking as compared with the past. Decently disposing of and burying the war dead in identifiable ways was a social demand that governments could not escape. People cared about the dead, their individuality, and their final disposal. In the aftermath of the Civil War, more than ever before, Americans were not willing to accept the disregard that characterized the policy the authorities had followed until then. Even grave markers had to be adapted to the new sensibility. As Union General Dana observed in 1866:

> Public opinion seems to be turning to a more permanent mode of marking the graves than by wooden head-boards, and I would respectfully give it as my opinion that the sentiment of the nation will not only sustain the expense of marble or other permanent memorial, but, moreover, that it will be likely to demand it in a few years, if not now established.
>
> (in Steere 1948: 160–1)

In accordance with the prevalent attitudes towards death in society, nineteenth-century Americans displayed an unprecedented interest in a respectable burial for their fallen relatives. Governmental agencies, hospital nurses, members of the Sanitary Commission, and the Christian Commission tried their best to provide proper, decent entombments. Such a policy cost an average of $9.75 a body and a total of $4,000,306 for the Federal soldiers identified and buried in the seventy-four national cemeteries (Faust 2008: 235–36). Overall, in the decade following the end of the conflict, 300,000 corpses were reinterred. Although, as already mentioned, no administrative machine for the disposal of dead soldiers was in place in the early years of the war and, as a result, most bodies remained unidentified, such 'a consecration of a nation's power and resources to a sentiment', Quartermaster Edmund B. Whitman rightly declared, 'the world has never witnessed' (in Poole 2009: 72).

Limits to the Democracy of Death in the Civil War

The emerging democracy of death, however, did not completely erase social differences between officers and enlisted men. In the Civil War, too, officers had privileged treatment over the average soldier. Their bodies were generally placed in metal coffins and then shipped back to their families. At the Arlington National Cemetery, Union officers were also provided with larger headstones than ordinary soldiers. As a Texas soldier observed: 'The officers get the honor, you get nothing. They get a monument, you get a hole in the ground and no coffin' (Faust 2008: 80).

Furthermore, racism affected the removal of the dead and the return of bodies. During the Civil War, it was common to call a truce for the exchange of dead officers, but their bodies could not be returned if they were black. Even the reburial programme of the Union discriminated between races. African American soldiers were buried in sections termed 'colored', but only a third of them were identified, as against a total rate of identification for the Union army of 54 per cent (Faust 2008: 236). Despite their efforts and sufferings in battle, the location of cemeteries for African Americans was symbolic of their place in society when alive: racial discrimination in life carried over into death.

Finally, it should be noted that the reburial policy included only the fallen for the Union. Defeated Confederates were not granted the decent treatment the federal government accorded to Union combatants. The Civil War national cemetery, as John Neff (2004: 132) pointed out, 'is best thought of as a synonym for *Union cemetery*'. Since Confederates did not deserve equal treatment in death, national cemeteries usually excluded their corpses or marked their graves with the label 'rebel'. For example, the Confederates who had fallen in the battle of Gettysburg were not buried individually or as decently as Union soldiers. Their remains were placed in a mass burial site outside the cemetery.[19] However, a number of private individuals and groups, especially women's associations like the Ladies' Memorial Associations, started a reburial programme for their war dead, which mobilized the white population of the South for the retrieval of bodies from Union cemeteries and reburial into Southern ground (Blair 2004; Janney 2008). The McGavock Confederate Cemetery in Franklin, the largest privately held Confederate cemetery in the United States, was created in 1866 on a two-acre property donated by the McGavock family

[19] Only under the presidency of William McKinley did the federal government decide to grant the Confederates a plot in the Arlington National Cemetery. Later, President William Howard Taft allowed the construction of a Confederate memorial in the same grave site (Piehler 1995: 66). Through these symbolic acts, the federal government finally granted the South a right of national commemoration.

and has been managed by the United Daughters of the Confederacy since 1905. The story illustrates how, even in the absence of the state, a local community was ready and able to cooperate in order to provide its fallen with appropriate burial and identification.[20]

The New and the Old Social Bodies of Soldiers: Individualization of Death and Glorification of War

Suggesting, as I have done so far, that the changes in burial practices concerning combatants and the creation of military cemeteries reflected social demands does not equate to contending that war commemoration was simply a mirror of civilian mourning practices. Although the individualization of burials manifested a societal need for mourning the war dead, the process was framed within the narrative of the nation. Through specific funerary rituals and commemorative functions, the fallen were turned into martyrs whose death took on a noble, political meaning. In fact, American Civil War cemeteries embody both a type of commemoration centred on the individual and a narrative centred on the collective and political meaning of war. Military burial grounds were not simply meant as places for the disposal of corpses and mourning; they were also supposed to teach the living about fallen heroes.[21] The way they were shaped was determined not only by the material and mortuary functions that they were to fulfil, but rather by the social and political meaning attached to the fallen.

Military cemeteries were a political statement that shifted the focus of war towards the soldier and his sacrifice; individual loss was strictly interpreted as a collective immolation for the common good. Indeed, while the individuality of the dead soldier had now achieved public acknowledgement, one should bear in mind that it still remained secondary to the holistic theme of the nation. Gettysburg, in particular, was built as a site of national military worship rather than a site of mourning for single individuals: here soldiers were buried state by state, with unknown soldiers placed in a different area.[22] Gettysburg was 'in form and conception designed to commemorate not the individual, but

[20] In the McGavock Confederate Cemetery, 780 Confederate soldiers were identified and 558 corpses remained unknown.

[21] For a different opinion on the American Civil War, see Mosse (1979: 8), who argues that war cemeteries were merely burial places. Such a view seems to stem from Mosse's belief that the first systematic cult of the fallen soldier occurred with the First World War.

[22] When the reburial policy at Gettysburg came to an end in March 1964, 3,354 soldiers were reinterred: 1,664 still remained unknown by name and 979 unknown both by name and by state.

the nation', historian Susan-Mary Grant (2005: 513) points out.[23] Arlington, America's most famous cemetery established on the property confiscated from the family of General Robert E. Lee, also fulfilled this political function.

What before had been merely a large amount of human flesh left to rot was now something to be retrieved and treated with care, something pious that the nation should honour and glorify. Lincoln's famous Gettysburg Address on 19 November 1863, which appealed to the principle of equality by defining the United States as 'a government of the people, by the people, and for the people', was delivered during the inauguration of the Soldiers' National Cemetery, just as recovery and reinterment of the bodies of the fallen was taking place. In his brief but potent speech, Lincoln (2013: 191) declared: 'We have come to dedicate a portion of that field, as a final resting place for those who here gave their lives that that nation might live.' As Garry Wills (2006) noted, Lincoln fashioned his dedication on the model of Pericles' funeral oration, which was meant to regenerate the nation through the bodies of the fallen. In other words, war cemeteries were intended not only as decent places of interment but also as cultural institutions conveying a political, didactic message to future generations, a point that was emphasized by the former general and future president James A. Garfield during the first Decoration Day (30 May 1868) ceremony at Arlington, when he stated that for 'love of country' soldiers 'accepted death' and 'made immortal their patriotism and their virtue' (in Underhill-Cady 2001: 62).

Military Cemeteries and Ossuaries in Europe

A few years after the end of the American Civil War, as a result of the Franco-Prussian conflict, Prussia established its first military cemetery (Grant 2005: 510). During the German wars of liberation against Napoleon, a few cemeteries 'were made here and there', but there was no systematic policy for the disposal of the war dead. For, as 'late as the Franco-Prussian war the bodies of soldiers were usually left to decay where they had fallen, or their bones collected in a nearby charnel-house' (Mosse 1979: 8). As of the beginning of the Franco-Prussian War, it was the non-democratic Prussia that planned to provide a decent disposal for its fallen men. Unlike France, the Prussian army provided soldiers with identity tags worn around the neck in order to provide accurate

[23] On the priority of the nation over individuals in Lincoln's Gettysburg Address, see also Nudelman (2004: 37).

medical care (Swain 1970: 512) and identify the dead more easily (Elliot 1999: 232). During the war, soldiers' burials were marked with improvised crosses, and, when known, the names of the fallen were written on paper and stuck on crosses with sealing wax. Moreover, at the end of the conflict, the Treaty of Frankfurt (1871), Art. 16, established mutual respect and maintenance of soldiers' graves.

Despite these policies, a large number of soldiers remained unidentified, which was especially true for the French, given the unpreparedness of their government for disposal of corpses. When it was impossible to identify the war dead, ossuaries (collection sites for bones) were built in their honour. At Metz, for example, the ossuary contains the remains of 7,636 soldiers, but only 178 have an individual grave. The charnel-house at Bazeilles near Sedan, housing both French and German remains, and the one at Loigny, which was built by a private committee (Becker 1997: 666; Varley 2008: 164) with a church known as 'The Battle', are some of the most famous military ossuaries of the Franco-Prussian War.[24]

Like Civil War cemeteries, military ossuaries were not only designed to offer peaceful resting spots for the unidentified remains of the fallen and to provide mourning places for grieving families. They were also intended as physical constructions with strong political connotations meant to foster a particular reading of the war. Despite their gruesome raw material, the message was not morbid but didactic. By putting the collected bones of the fallen on public display, ossuaries sought to glorify the fallen soldiers and the war (Becker 1992). Such buildings placed dead soldiers within a long tradition of sanctuaries glorifying the relics of important lives.

Care for the graves of the dead was also conspicuous during the Second Boer War (1899–1902). Locating and recording military graves became a task formally assigned to the Army itself—namely, to the Royal Engineers. The Guild of Loyal Women, a voluntary organization, helped the army with compilation of registers and the marking of burials with iron crosses (Heyningen and Merrett 2002). Although remarkable, their work was not followed by any appropriate, properly funded maintenance policy in the Transvaal region. The final result was a disappointing and dilapidated system of war cemeteries that largely failed to honour the war dead (Summers 2010: 11).

[24] Two monumental war ossuaries were built at Solferino and San Martino in 1870, eleven years after the battle had taken place. As it was impossible to identify the remains of the dead, the corpses were exhumed, and the bones and skulls collected into two ossuaries.

The First World War and the Disposal of Soldier Remains

Despite many shortcomings, by the eve of the First World War the practice of burying soldiers in military cemeteries had, therefore, already been established. Unsurprisingly, finding soldiers' corpses and interring their remains became a major issue in the countries who fought in the First World War. Several war cemeteries were built in Europe, especially in the regions where great battles were fought, such as northern France and Flanders. Battlefields such as the Marne, Ypres, the Somme, and Verdun were transformed into sanctified resting places for the fallen. The ossuary at Douaumont, near Verdun, is only the best known among a multitude of such sacred burial sites.

Right from the start of the war, Britain established a set of policies for the disposal of the dead. In 1914, Secretary of State for War Lord Kitchener created a mobile unit tasked with seeking, identifying, and marking the corpses of the fallen. One year later, the government bought French land for the permanent burial of British soldiers. By March 1915, the Red Cross units in charge of body disposal were reorganized as the Graves Registration Commission. In 1916, as the number of casualties mounted, combatants were buried where they fell, but a year later a number of measures were implemented to simplify the concentration and identification of the war dead from Britain, the Dominions, and the Empire. It was one of the main goals of these policies to record the names of the fallen and preserve their identities.

Initially, relatives' requests for traditional civilian funeral rites were not part of the agendas of the French political and military leaderships. At the beginning of the war, the French directories for the burial of enlisted men proposed using common graves, which they regarded as an appropriate way of dealing with the bodies of the war dead. However, French society was outraged at the idea and stood out against it. The disposal of soldiers' bodies was to conform to civilian practices in time of peace, which meant at least that corpses should be placed in single coffins after identification.[25] Only after 'bitter negotiations between the society at large and the civil and military authorities' (Capdevila and Voldman 2006: 46) did the government meet the citizens' demands.

As in the previous epoch, however, mass graves had to be used in the First World War, especially when the dead were interred by their surviving comrades (Wilson 2008: 151–4). Just as in the American Civil War, burying the

[25] An interesting French exception is to be found at Laurens. Here, in 1923, sixteen bodies of men killed in war were buried at their families' request in a communal grave over which a monument was erected. But even in this case, parents personalized the burial by inscribing the names of the deceased and by placing photographs (Capdevila and Voldman 2006: 181).

fallen with civilian rites often proved an unfeasible luxury. Before a major battle was fought, mass graves were dug. When the dead could not be interred, sometimes for days and even for months, they started to rot and stink, as in times past. For instance, at the Somme in February 1917, when the Germans retreated to a new line, the British were able to move into No Man's Land and collect the corpses of their fallen, which had been lying in the open for eight months. Although they were decomposing, most of the bodies could be identified by identity discs and given proper burial (Middlebrook 1984: 297).[26] But in the First World War *fosses communes* were considered temporary gravesites, and the survivors were expected to handle each corpse as carefully as possible. Cadavers interred in temporary burials were expected to be relocated to proper military cemeteries at the end of hostilities. The policy of concentrating corpses from isolated grave sites was followed by virtually all warring nations. Moreover, individual tombs gradually replaced communal graves thanks to improvements in military burial services.

The issue of military cemeteries figured prominently in the post-war period as well and, significantly, received some space in the Treaty of Versailles (sect. II, 'Graves'), Art. 225 of which specifies that 'Allied and Associated Governments and the German Government will cause to be respected and maintained the graves of soldiers and sailors buried in their respective territories'. The rule was observed even by local communities in France that felt insulted by the presence of enemy bodies in their soil. In most warring nations, the search for and identification of dead soldiers continued long after the conflict had ended. Corpses were exhumed, then named, and, when feasible, reinterred individually. Although these burial policies appeared to be important everywhere, their implementation varied from country to country.

In April 1915, the British banned repatriation of their dead at least for the duration of hostilities (Dendooven 2014: 66–7). The policy was confirmed in the summer of 1919 when the British authorities decided, respectively, to forbid exhumation of their soldiers' remains, to build war cemeteries abroad, and to erect war memorials at home, where relatives could mourn their loved ones (Longworth 2003: 14). The British government met the people's demands for appropriate, individualized graves, and sent hundreds of thousands of

[26] In virtually all European countries, First World War soldiers were provided with identification tags or discs, which might be made of metal, cellulose, or boiled leather. The British, for example, used nine different types of discs. Systematic organization of identification procedures would start only after the war. The Geneva Convention for the Amelioration of the Condition of the Wounded and Sick of 1929, under Art. 4, introduced a standardization process for identification consisting of an identity disc with a detachable part to be passed to authorities.

gravestones to France after the war's end.[27] In the 1920s, 30,000 British bodies were reinterred, even though only a quarter could be identified, since their fibre identity discs had rotted (Winter 1978: 260). In total, the British were unable to find or identify 517,000 soldiers out of the approximately one million they had lost in the war. Indeed, the rudimentary identification method meant that the whole process was inaccurate at best (Meigs 1994: 144).

The fact that the bodies of the fallen in First World War cemeteries were sometimes buried in communal graves with headstones arranged in rows to provide the false impression that each man had his own individual burial (Fussell 2013: 6) tells something about how the reality of anonymous death in war needed to be reconciled with the social requirements of a proper, individualized grave.

Unlike the UK, the United States arranged for its fallen soldiers to be exhumed and shipped back home if their families so wished. Those who were not reclaimed and buried in Allied soil were concentrated in eight permanent cemeteries located in France (six), Belgium (one), and England (one) under the direction of the American Battle Monuments Commission; those interred in Russia or in enemy territories such as Germany and the former Austrian-Hungarian Empire were repatriated. These operations lasted from 1919 to 1922 and cost the American government over $30 million (Budreau 2010: 21). According to Anders (1988), 47,000 bodies were returned to the United States and 30,000 laid to rest in Europe. Almost half of those who had been found in Europe were buried in the Meuse-Argonne cemetery with no distinction of race or rank. In 1929, Congress decided to finance pilgrimages to the grave sites in Europe for the families whose loved ones had remained there.[28] Unlike other warring countries, the United States was able to identify almost all its dead: only 3,350 soldiers were not found, or were recovered but unnamed.

Burial policies in France were far more complicated. In 1915, the French government decided to finance the creation of a series of military cemeteries for the fallen, but during the war corpses were also buried in the so-called Zone-Rouge of civilian cemeteries. While during the war the French government had forbidden exhumation of soldiers' remains (Prost 2011), in 1920—after a burning, painful public debate, and a period of widespread illegal exhumation (Sherman 1998: 451) carried out by the so-called *merchants of*

[27] Despite the egalitarian attitude of British burial and commemorative policies, outside Europe the Imperial War Graves Commission discriminated on the basis of 'race and creed'. For example, the identified graves of Africans were 'abandoned and the names reclassified as "missing"' (Barrett 2014: 82). At best, the sacrifice of these troops was honoured collectively.

[28] For obvious reasons, Germany was not able to exhume and transmit the corpses of its soldiers from the areas it had occupied during the war (Winter 1995: 27).

death (Dendooven 2014: 69)—the government met the families' requests and recognized their right to claim the bodies of their loved ones at state expense, a policy funded by the Special Service for the Repatriation of Bodies under the leadership of André Maginot.[29] By the beginning of 1923 the process was finally completed, with 240,000 coffins returned to the families—that is, 34 per cent of the approximately 700,000 identified fallen soldiers whose repatriation to relatives was approved (Smith et al. 2003: 73).

In Italy, too, the issue of military burials became prominent in the aftermath of the war. A National Committee for the Honorary Burial of Italian and Allied Fallen was created in 1919 (Dogliani 1999). Then, in January 1920, the responsibility to trace and exhume the bodies of the war dead was assigned to the Central Office for the Care and Honour of the Fallen.[30] In northern Italy many small military cemeteries were built after the war, where the remains of the fallen were moved from the countless temporary mass graves scattered near the front line. But in the early 1930s, displeased with the modesty of these burial places, the Fascist government launched a reburial programme with the construction of thirty-six monumental military cemeteries and ossuaries, such as the ones of Montegrappa, Pocol, Fagarè, Asiago, Caporetto, and Oslavia, where hundreds of thousands of corpses were concentrated (Malone 2019b: 201).[31] The *Sacrario militare di Redipuglia*, in Friuli, is the largest and most famous of these burial sites; here the remains of over 100,000 Italian soldiers, 60,000 of whom were unknown, were laid to rest in two ossuaries and a monumental staircase shaped as a military formation with the tomb of the Duke of Aosta (commander of the Third Army) at its base. Even in the case of these monumental cemeteries and ossuaries, where the concentration of the war dead was meant to symbolize the power of the new Fascist regime and its superiority over the citizens (Dogliani 1996: 382) the names of the known war dead were nevertheless listed in alphabetical order.[32]

[29] The French government, while recognizing the right to claim exhumation and reinterment at public expense, provided economic incentives to those families who decided to leave their loved ones where they had fallen (Sherman 1998: 453).

[30] On the figures of the Italian war dead, see Bregantin (2010).

[31] About 50,000 soldiers' corpses were returned to the families (Bregantin and Brienza 2015: 42).

[32] Although in the area of military cemeteries there are certainly important differences between the policy of Fascist Italy and, for example, the British and French approaches, I believe Antoine Prost (2011: 41–2) overemphasizes the extent of such variations. First, like the Tyne Cot Cemetery, in Redipuglia, the individual names of the fallen are listed. Secondly, like Redipuglia, the Ossuary of Douaumont is a colossal white architecture. Moreover, this French ossuary has a 'belligerent' structure, too: it is shaped as a shell or, according to some, as a sword. Finally, similarly to Redipuglia, under the main entrance of the Douaumont ossuary there is the tomb belonging to a military commander—General Anselin, who fell in battle on 24 October 1916.

From Great Commanders to Commoners: The Rise of Commemoration for Ordinary Soldiers

It is now received opinion that the First World War was the time when the modern commemoration of the fallen was born. However, this widely held conviction does not withstand historical scrutiny. Construction of war memorials dedicated to ordinary soldiers was not a result of the shocking tragedy of the First World War but started before that conflict. The First World War was the peak of commemorative practices that had been emerging since the late eighteenth century, had gradually developed in the second half of the nineteenth century, and finally boomed during the First World War and its aftermath.

One of the earliest examples of modern commemoration appears to be the *Hessendenkmal* in Frankfurt, built in 1793 to celebrate the Prussian fighters who died on 2 December 1792 in the successful attempt to drive back French revolutionary forces. Not only were the names of officers killed in battle inscribed on the monument, but also those of ordinary soldiers. Apart from that and a few other eighteenth-century monuments,[33] it is in the second half of the nineteenth century, especially in the last thirty years, that war memorials dedicated to ordinary soldiers were erected in Europe and in the United States alike. In both continents, statues of anonymous soldiers began to appear near battlefields and in town squares. In the process, the social bodies of soldiers began to transcend the old social distinctions.

The American Civil War, which had developed military cemeteries, was innovative in the area of memory as well. Although the pantheon of heroes of the Civil War was not completely open to the average soldier—Lincoln, Grant, and other commanders were the subjects of most memorials around the country—countless numbers of statues commemorating the service of ordinary soldiers were erected in the aftermath of the conflict. In the last third of the century, especially in the 1890s (T. B. Smith 2008), the sculptural figure of the common soldier emerged as the most common device to commemorate the war. Such memorials, especially at a local level, emphasized not only military victory, but also the human cost of war, and in some instances listed the names of the community's fallen (Baruch and Beckman 1978).[34] The basic

[33] For another few examples, see Mosse (1990: 47).

[34] For a catalogue on war monuments in the United States, see the Smithsonian's Inventory of American Sculpture.

formula for these monuments was a single infantryman standing on a column or pedestal, rifle in hand.[35] In some memorials, especially those erected on battlefields, the soldier figure poses as if in active combat (Savage 1997: 164, 248). While in the North many of the monuments for common soldiers were erected by the main veterans' organizations (for example, the Grand Army of the Republic) and its lady auxiliaries, in the South they were financed mainly by women's organizations (Savage 1997: 247), but their shape remained similar.

As a clear political statement, the Germans erected two hundred monuments in Alsace Lorraine in the wake of the Franco-Prussian War to commemorate their soldiers' sacrifice (Becker 1997: 660). Within a mere seven years of that war's end, in a struggle over memory that can truly be described as a 'war fought with other means', the French responded by building 457 memorials in honour of their war dead and many others in the following years (Varley 2008: 104). Examples of these monuments include Louis-Ernest Barrias's *The Defense of Paris* and the splendid tomb of the National Guardsmen in Colmar, where a French soldier rises from the grave. The statue by Frédéric Bogino in Mars-la-Tour, which features an allegory of France supporting a wounded soldier, combines previous themes such as the nation with the new one of the ordinary fallen combatant. Even where the war dead are secondary to other themes such as *la patrie* or the Republic (often symbolized by the image of Marianne), it is significant that the memorial is intended to represent the war effort through fallen soldiers.

Memorials were erected not only in cities, cemeteries, and on or near battlefields but above all in hundreds of the villages from which the fallen had hailed. In fact, most of the French monuments were built at the behest of individual communes rather than the state or the army (Troyansky 1987: 122; Varley 2008: 15) and support by such organizations as *Le Souvenir français* (Prost 1997: 308), founded in 1887 with the explicit goal of commemorating the French dead from all armed conflicts. In fact, the Paris government played only a minor role in the erection of local monuments and in organized events of commemoration. Honouring the dead was largely a product of local initiative and funding, which did not end with the creation of memorials, but was perpetuated in regular annual gatherings. Although local communities

[35] While women were also, though rarely, commemorated in these monuments, the African American soldier was virtually absent. Although there are monuments dedicated to the issue of emancipation—with Lincoln at their centre and/or with a kneeling black slave—in his study on monuments in nineteenth-century America Kirk Savage finds only one African American standing-soldier monument, which is located in Norfolk, Virginia. Not surprisingly, Savage (1997: 208). concludes that, for 'all its novelty, the common-soldier monument was "fundamentally reactionary"'.

and groups differed greatly in terms of political goals and cultural agendas,[36] it is telling for this research that they should share the cult of the fallen as a leading theme of war commemoration. As Varley (2008: 18) argued, the 'post-1871 cult of the fallen placed unprecedented emphasis on the mass of common soldiers, invoking their patriotic self-sacrifice to lift and unify the nation after its collapse'. It was now the rank-and-file soldiers who were commemorated and celebrated, not only their generals and officers.

Unlike future memorials, most of these monuments bear inscriptions not of the names of ordinary soldiers killed in battle, but rather of the names of the regiments that were engaged in combat (Becker 1997: 661) and, obviously enough, those of the fallen officers, as in the Bazeilles monument. A few, however, also list the names of common soldiers, which is equally true in the case of Prussia. In the city of Marburg, for example, the names of the twenty-five men fallen during the war are carved on a monument along with those of their home towns and units. In the cemetery of Mühlacker, there is an obelisk in memory of the Franco-Prussian War, with three panels listing seventy-five names at the foot of the stone. Although the names do not all belong to war dead—only a few are followed by a cross—the list is a visible testimony to a remarkable change in the commemoration of war.

Further examples of the new attitude towards commemorating common soldiers are to be found in the late-nineteenth-century memorials erected by European armies to remember the dead of small-scale colonial wars. For instance, a monument dedicated to 548 Italian soldiers killed at the battle of Dogali (Eritrea) was placed in front of the railway station in Rome in 1887 with four bronze plaques displaying the names of the fallen soldiers.[37] In Britain as well, a great burst of memorial construction occurred after the Boer War, which resulted in hundreds of monuments being erected in many British cities and villages, such as Eamon Bridge in Cumbria, where the two villagers killed in war are portrayed (Borg 1991: 94) or on Coombe Hill, near Wendover, where a plaque lists the names of the fallen from Buckinghamshire.[38] It should be noted that commemoration of the rank-and-file soldier in colonial wars began when European armies were largely formed of a

[36] Republicans and the local population honoured their dead soldiers, the Left commemorated the Communards killed in the spring of 1871, and the Church too 'became involved, as the basilica of Sacré-Cœur in Paris was begun in 1876 as an act of expiation' (Troyansky 1986b: 351–2).

[37] While the square is still named Piazza dei Cinquecento (Square of the Five Hundred), the memorial is now located in Viale Luigi Einaudi between Piazza della Repubblica and Piazza dei Cinquecento.

[38] At the beginning of his study on the Crimean War, historian Orlando Figes (2010) gives two examples of local memorials in Britain (Witchampton, Dorset) and France (Héricourt) dedicated to the fallen of the Crimean War. For other examples, see King (1998: 20–44).

handful of professional soldiers who fought in relatively low-casualty imperial military interventions.[39]

With names now inscribed on monuments and plaques, commemoration ceased to be a generalized statement about war and victory. By displaying individual names, states recognized that war was fought not by a collective entity, but rather by individuals with personal identities: the ordinary soldier was now fully commemorated as a legitimate member of the national community. These long-neglected human beings were at last celebrated like the great heroes of the past. Though inscriptions of names are not always present, the very fact that they are now usually used highlights the shift from a message based on the traditional meaning of collective sacrifice to one that gives prominence to individual death. Thus, naming the dead was a major factor in granting individual soldiers unprecedented national and political significance.

War-memorial construction in the second half of the nineteenth century also shows that inclusion of common soldiers in the commemoration enterprise was not an effect of the lethality and mass death of the First World War,[40] but rather the product of a slow movement towards recollection of those who bear the major brunt of war. There was a shift from a cult centring on great heroes to the celebration of the common man, embodied in a variety of commemorative material. Memorials dedicated to ordinary soldiers before 1914 were less numerous than First World War monuments, which might be explained by the simple fact that the casualties were so much lower, except that, again, inclusion of the average soldier in commemorative practices antedated the First World War.

Whereas it has been shown that the origins of the cult of the fallen date back to the second half of the nineteenth century, the decade after the First World War, to borrow Pierre Nora's phrase, can truly be described as an *age of commemoration*. As memorialization of the conflict came to include all previous developments regarding the commemoration of ordinary soldiers, the following section will give a detailed picture of the memorial practices of the First World War.

[39] For instance, 22,000 British soldiers lost their lives in the Boer War, approximately the same casualty toll as on the first day of the Somme offensive.

[40] David Cannadine (1981) saw the scale of death of the First World War as originating a change in attitudes towards death. For an interpretation stressing the role of the First World War as a turning-point in attitudes towards military casualties, see Jalland (2010). However, it was not, surely, the sheer number of the war dead in itself that sparked the modern cult of the fallen; for losses in previous great conflicts such as the Thirty Years' War and the Napoleonic Wars had not triggered anything like the commemorative practices that came into being in the context of the First World War.

First World War Commemoration

A. J. P. Taylor (1980: 140) famously wrote that 'idealism perished on the Somme'. In similar vein, Paul Fussell's 1975 classic *The Great War and Modern Memory* discerned a profound fracture between the chivalric romantic idiom of past wars and the disenchantment and cynical disillusionment of the First World War, which, in the opinion of that eminent cultural and literary critic, was primarily a consequence of the meaningless slaughter of the trenches. In Fussell's view, the main cultural product of the First World War was irony, conveyed by soldiers in the cynical, sarcastic, and vitriolic words of their diaries and letters, and in the artistic works of writers and poets who fought in the conflict, such as Edmund Blunden, Robert Graves, Siegfried Sassoon, and Wilfred Owen. The last-named cited Horace's *Dulce et decorum est pro patria mori* to satirize 'the old Lie' of dying for the glory of one's country.[41] According to Fussell, death in war came to be construed as a useless waste of lives rather than as an act of love and heroism for the nation.

There is, of course, more than a grain of truth in Taylor's and Fussell's views, but they are more applicable to how the war was perceived, understood, and told by soldiers, rather than to the meaning attached to it by governments and societies.[42] There was, indeed, no irony in how states and societies celebrated and remembered the war effort; the sense of absurdity at a personal level was coupled with a grave, solemn process of commemoration in the public sphere, which began in the initial phases of the war. For, as early as January 1915, David Lloyd George, the British prime minister for much of the war (December 1916–October 1922), declared that the British army was 'a force of a totally different character from that which has hitherto left these shores. It has been drawn almost exclusively from the better class of artisan, the upper and lower middle classes ... the people of this country will take an intimate and personal interest in its fate' (in Lambert 2012: 306).

Although the way in which British soldiers were expended on the battlefield, as we shall see, did not prove any special interest by the leaders in the fate of their men, it is certainly true that the fallen of the First World War were

[41] For a study highlighting the chasm between political rhetoric and personal expression of grief for the French fallen, see Trevisan (2001).

[42] Fussell's *The Great War and Modern Memory* has also been criticized on account of his biased selection of material for analysis, especially the fact that he relied mostly on evidence coming from the affluent part of British society (Winter 2013: p. xii). Moreover, in his study on letters and diaries by Australian soldiers, Bill Gammage (1974) has shown that for some of them the wartime years were the happiest of their lives. This is particularly significant considering that Australian armed forces suffered the highest percentage of dead and wounded among Allied countries. For similar arguments concerning the First World War and other conflicts, see Bourke (1999) and Hart (2008: 12).

honoured and commemorated with an intensity never seen before. During the war and its aftermath states spent a vast amount of resources on commemoration. According to Alan Borg (1991: pp. ix–x), memorial construction in the wake of the First World War represents the greatest communal arts project ever undertaken, what he termed 'the classic era of memorial construction'. Such a definition is not only justified for Britain, but applies to all combatant nations. For example, from 1920 to 1925 over 30,000 war memorials were erected in France, reaching the outstanding number of 176,000 in the following years (Smith et al. 2003: 160).

Commemoration in Britain began while the war was still on, when temporary memorials (for example, street shrines), financed by a variety of actors such as clergy, benefactors, and resident groups, were erected in city neighbourhoods in honour of the local fallen (Connelly 2002: ch. 2, 'War Shrines').[43] The street shrines were explicitly meant to represent, in the words of Alex King (1998: 55), 'the idea that citizens in arms were bearers of special moral worth and that their sufferings were achievements of public significance'. Likewise, memorial construction in France emerged during the war and was often the outcome of private citizens cooperating with local and national authorities (Prost 1997: 308). However, the age of commemoration actually began when the guns fell silent, when a multitude of rural and urban locations were turned into what the eminent French historian Pierre Nora (1984) called *lieux de mémoire* (sites of memory). Even in the United States, which paid a relatively low cost in terms of human lives in comparison to the Civil War, memorial production boomed after the conflict, and massive monuments like *Wars of America* (1926) were erected honouring the fallen.

The type and shape of memorials varied greatly—ranging from hospitals to libraries, public halls and scholarships, but the most favoured format was the monument (Borg 1991: 69), which could be built in cities, villages, or on battlefields where armed clashes had taken place. A number of memorials were experimental, others were designed in conventional forms; some were small and cheap, others gigantic and highly expensive. Although memorialization took several forms according to national and local cultures,[44] it should be noted from the outset that the practices commemorating the First World War have a common history with few national boundaries (Winter 1995: 227; Smith et al. 2003: 164).

[43] In France, too, commemorative rites and memorial construction began while the war was on (Prost 1997).

[44] For example, war memorials in Australia record not only the names of fallen soldiers but also those of other combatants, whether they were killed in war or not (Garton 2006: 224).

Arguments for consistency in the commemorative response by different societies to the mass carnage of the First World War are, no doubt, fraught with risk because of the need to generalize. One might reasonably wonder how similar commemoration practices could be shared by countries that won or lost the war. How could the same practices suit both victory and defeat? The answer, as we shall see, lies in the evidence itself.

Four main common features of First World War memorials speak significantly for attitudes towards death in war: the human sacrifice of common soldiers is commemorated, memory is construed through the use of Christian metaphors, especially as a martyrdom for the nation in the image of the Passion of Christ, the naming of soldiers plays a central part in the memorialization process, and, finally, the cultural invention of the Unknown Soldier (or Unknown Warrior in Britain) enjoys widespread popularity.

The Sacrifice of Common Soldiers

Whereas the central commemorative theme used to be triumph in war revolving around military and political leaders, in the aftermath of the First World War both victory and defeat were honoured through the representation of the war dead, with the exception of the Soviet Union, which nevertheless celebrated the fallen of the Revolution. While the state had always created and revered its consecrated heroes, the ranks of holy ancestors were now entirely open to humble and obscure combatants. The invention of the Unknown Soldier, memorial days, and the countless national and local war cenotaphs dotting the urban and rural landscape of Europe attest that the failure to acknowledge the soldier's role in war was now a thing of the past.

In the emergency of the First World War, it might sound natural that states should use the language of sacrifice and glory in order to increase enlistment and motivate their combatants. The revival of this language, however, was not a simple repetition of long-established warrior ideals, but rather a departure from tradition, as the sacrifice of ordinary soldiers was now publicly recognized. Trench warfare no doubt lacked the glory of chivalric combat, but death on the battlefield became something glorious and noble likewise for those common combatants who had figured as anonymous, insignificant automatons in the early modern period.

At the same time, fewer monuments than in the past were erected to celebrate military commanders. Great generals and war leaders were still honoured, largely by means of statues, but their number was small in comparison to the memorials dedicated to the deceased rank and file, which seems to support the view that—at least in the area of commemoration—plain soldiers had become as important as the famous dying heroes of the past.[45]

The War Dead as Martyrs

The way in which the sacrifice of soldiers was represented is itself meaningful. States portrayed the death of common soldiers as a martyrdom for the nation construed in the image of Christ's sacrifice, as the triumph of the spirit over the flesh. Just like Jesus, who had died to rescue mankind from hell, the fallen soldier had given his life in the patriotic act of saving his nation. Jesus' bloody ordeal marked the origin and foundations of Christianity. Likewise, offering one's blood in combat became the emblem of national unity in the First World War. The Christian narrative of salvation, blending death (Passion) with victory (Resurrection), provided a perfect commemorative symbol for the First World War (Mosse 1979: 5).

The association of death and religion is far from surprising, since the glorification of death is a founding pillar of Christianity. 'Death was the outcome of sin, yet sin was destroyed by Christ's pure life outpoured in his atoning death' (Davies 2005: 7). Although the dead soldier could not rise again, it was the country, like Christ himself, that would rise. Thus, the dead body of the soldier came to symbolize the nation conceived as a time-extended entity with the power to elevate the soldier above the limitations imposed by mortality. Dying in battle was portrayed as a sacred, redemptive act that would give fresh blood to the people. In the famous monuments *Terre de France*, sculptured by Maxime Real del Sarte, we see the grave of a soldier from which sheaves of wheat are germinating in great number. According to Deborah Buffton (2005: 28), the message is that 'the blood of the dead soldiers brings forth new life to reinvigorate the country'. Or, as a French soldier wrote, the war dead had 'given their blood that France may live' (in Watson and Porter 2008: 148). As a result, the nation became—in the famous phrase coined by President Raymond Poincaré—the *union sacrée*. The sacrifice of blood becomes a beautiful act of love for the polity, and death is accorded a positive meaning in the history of the nation, to the extent that the butchery of Verdun is framed by Poincaré within an idiom of beauty and morality: Verdun 'represents … all that is purest,

[45] Traditional forms also included military posters and other small artefacts, such as the well-known *images d'Épinal* in France, which were turned into an instrument of patriotic propaganda during the war with a view to sanitizing the effects of the conflict.

best, and most beautiful in the soul of France. It has become synonymous with patriotism, bravery, and generosity' (in Prost 2002: 50). As in the Christian message, death in battle was portrayed as a necessary sacrifice from which a new life might spring.[46] In the past, it was the shedding of the enemy's blood that purified the spirit of a community, but now that purgation was performed by the blood of one's own citizens.

By way of constructing an equivalence with the Passion of Christ, commemorative sculpture relied on the *Pietà* and on the image of a soldier's body in the Madonna's arms. Christian iconography and the lexicon of the Bible were also employed in other ways: the war dead are attended by angels (Winter 1995: 91), or it is Jesus himself who cradles a dying soldier, as in the fresco displayed in the chapel of the Italian *sacrario* at Redipuglia. In French monuments the nation is often portrayed as a woman in the likeness of the Virgin Mary, which turns the war memorial into a representation of the 'Catholic prayer *Stabat mater dolorosa*, the holy mother of sorrows' (Smith at al. 2003: 162). The sentence from the Book of Ecclesiasticus (44:14), 'Their name liveth for evermore', chosen by Rudyard Kipling for British and Imperial war cemeteries, is another example of the connection between Christianity and war that was typical of the commemorative practices of the First World War. Christian metaphors and the process of re-sacralization are also prominent in several First World War poems and novels, even in the works of those authors who displayed an ironic attitude to the war, such as Sassoon's *The Redeemer* and Leonard Green's *In Hospital* (Fussell 2013: 128–9).

The theme of sacrificial death was again chosen by the American Battle Monuments Commission for the design and construction of eleven memorials and cemeteries in Europe. Use of the marble cross as a grave marker instead of the traditional rectangular tombstone was specifically meant as a symbol of Christ's Passion (Robin 1992:, 120).[47]

Interestingly, even in a defeated country like Germany, the choice of these symbols was intended to justify the war effort and revive the potentially dying national spirit (Mosse 1990). The inscription 'They shall rise again' in German monuments, with a clear political message (Goebel 2004: 492–5), was

[46] Hence, the sacrificial ideology inscribed in the idea of a *beautiful death* was far from being only a brainchild of Fascism and Nazism: it was also a common western motif during and after the First World War.

[47] When the dead were Jews, the marble cross was replaced by a marble Star of David as a grave marker, but, to the disappointment of the Jewish community, the cross was maintained in memorial chapels (Piehler 1995: 101). The French used a dome-shaped tombstone for the graves of the Muslim dead.

likewise based on the Christian metaphor of resurrection and renewal.[48] Thus, the notion of sacrificial death in war transcended the political and military distinction between triumph and failure. In other words, similar commemorative forms were employed in the aftermath of the First World War to evoke opposite political realities.

By enveloping mass death in a solid Christian framework, states purposely and creatively transformed the destruction of millions of bodies on the battlefields into a regenerative event. However, the religious language and shapes should not be misread: the content was meant to sanctify the nation, the land, and the historical tradition, rather than the Christian God (Troyansky 1987: 130).[49] Although Christian and nationalist imperatives merged into a narrative of immortality, it was the latter that prevailed over the former. This sort of *militarized Christianity* was geared to proclaiming that the soldiers had not died in vain.[50]

Naming the Fallen

Another major feature of First World War memorials is the systematic listing of soldiers' names on monuments (Bushaway 1992). Along with the name and occasional biographical particulars, the dead man's life story is included in the collective narrative of the nation. Displaying the personal details of common soldiers appears to be a radical change from the typical war monuments of early modern Europe, where only officers enjoyed the privilege of having their names inscribed.

The size of name-inscribed memorials varied greatly. They include either small structures, like those erected in the fallen soldiers' birthplaces, or huge, architecturally imposing buildings. One of the most famous gigantic memorials is the Menin Gate at Ypres (Belgium), designed by Sir Reginald Blomfield, which bears the names of more than 54,000 men with no known graves. Another example is the Memorial to the Missing of the Somme on the Thiepval Spur, where German machine guns inflicted massive casualties on the first day of the offensive. The walls of the colossal red-brick building—designed by Edwin Lutyens—are carved with the names of over 70,000 men whose burial places are unknown.

[48] The most common grave marker that the visitor sees in a German First World War cemetery is, however, the rectangular tombstone.

[49] For a religious interpretation of this type of commemoration, see Becker (1989).

[50] Moreover, as Bruce Lincoln (1991) noted, the idea that creation requires destruction does not only belong to Christianity but is also a typical tenet of several Indo-European cultures.

The 'commemorative hyper-nominalism' (Laqueur 1994: 160) to be found in First World War memorials seems to have been affected by a more general strain of sensibility of a social nature, aptly summarized by Thomas Laqueur (1996: 127): 'everyone has a memorable life to live, or in any case the right to a life story.' The symbolic importance of listing the names was expressed, for example, by the mayor of the French village of Villerbon, who, during the dedication of the local war memorial, declared:

> The tribute to the unknown soldier in Paris, offered in the name of all France, is a tribute to all, but if history cannot preserve the names of those hundreds of thousands of brave men who fell in defense of the sacred soil of the fatherland, those who knew them, at least, must retain an imperishable memory of them. Their names must be synonymous with glory and abnegation, with patriotism and duty fulfilled, in the countryside where they were born and lived. Finally, future generations must be able in case of need to reinvigorate themselves through the reading of these valiant names.
>
> (in Sherman 1998: 443)

This commemorative practice appears the more impressive in that governments had to spend lavish amounts of financial, political, and administrative resources (Laqueur 1994: 155) in order to inscribe, record, and preserve the individual names of the dead.

The Invention of the Unknown Soldier: The One who Stands for All who Fell

The main commemorative innovation of the First World War is the Tomb of the Unknown Soldier/Warrior, an idea generally attributed either to Dean Ryle or to the Reverend David Railton, a military chaplain who served in Flanders with the British army.[51] After the project had been accepted, it was decided that the Unknown Warrior should be buried in Westminster Abbey in London on Armistice Day, 1920. Before being entombed, the body was carried across the country, and people paid homage to the unnamed combatant at each stop. Finally, his corpse was placed next to kings and queens. The same ceremony took place on the same day (11 November 1920) in Paris, where the *tombe du Soldat inconnu* was housed under the Arc de Triomphe.

Before then, the outpouring of grief had been strictly reserved for national heroes, but now there was a nameless soldier at the centre of public mourning.

[51] For a fuller review of these attributions, see Wittman (2011: 35), who argues that the idea was popular among the ranks before entering the public debate.

This anonymous figure replaced kings and generals in the heroic remembrance of war. In the following years, other unnamed soldiers were buried in some of the belligerent nations with massive celebrations gathering thousands of people in capitals such as Brussels, Lisbon, and Rome, and, in the United States, in the Arlington National Cemetery. In Italy, the Unknown Soldier was housed beneath the Altar of the Nation in the massive building of *il Vittoriano*, originally built to celebrate King Vittorio Emanuele II and unification of the country.

Thomas Laqueur (1994: 163) writes of the unknown soldier that, 'by being so intensely *a* body, it was *all* bodies'. His argument seems to contend that memorializing the unnamed soldier is not inconsistent with the naming practice but is actually part of it: 'the opposite end of the same discursive strategy that is evident in the enumeration of names' (Laqueur 1994: 163).

The grave of the Unknown Soldier was coupled with a cenotaph, a burial place that contains no body and therefore becomes the tomb of all the fallen.[52] Here, the absence of physical remains stands as an extreme example of the significance acquired by the social body of soldiers: so powerful is that social body, it does not even require the presence of its physical referent. The most popular cenotaph is Lutyens' in Whitehall, London, but more can be found in several other European and extra-European cities, villages, and battlefields. In London, a massive gathering of 400,000 people saluted the unveiling of the Cenotaph on 11 November 1920 (Greenberg 1989: 11), and many more gathered on the following days. It was originally devised as a temporary monument, but popular demand forced the government to create a permanent structure (Cannadine 1981: 221–3; Lloyd 1998: 50–63). In Paris, too, a monumental cenotaph with the inscription 'Aux Morts pour la Patrie' was placed under the Arc de Triomphe and was the pivot of the victory parade on 14 July 1919 (Smith et al. 2003: 171).

First World War Commemoration: Mourning Human Loss or Sacralizing the War Dead?

According to Jay Winter (1995), First World War memorials were erected primarily not to glorify the dead but rather to soothe the pain of those who had survived. Somehow, they were similar to cenotaphs in that they were empty

[52] The cenotaph, however, is not a modern invention. In the passage preceding *Pericles' Funerary Oration*, Thucydides (2.34) recalls the Athenian practice of using an empty bier to honour those soldiers whose corpses had not been found and returned to Athens.

graves meant to comfort grieving families. As Winter contends, war memorials were unquestionably sites of mourning and bereavement, since families who had lost their loved ones visited them on a regular basis; they provided a space for personal expression of grief, especially where there were no bodies to grieve over. The objects and rituals of remembrance cannot be reduced to mere mourning practices, though. Memorials were expressly designed for a specific political purpose. Governments tried to give a certain political meaning to their war effort by devising memorials with traditional, romantic images and inscriptions, which embody not only the themes of loss of life and bereavement, but also the political notions of nationalism, heroic patriotism, glory, and even race and imperialism in some British and French monuments.

First World War memorials are usually large in size, vertically erected, and realistic in their representations. They display national symbols and make explicit reference to the sacred war that was fought. Most of them celebrate ancestral warlike values, highlighting the nobility of dying and killing for the national cause. They are state-consecrated readings of history revolving around the ideal of military struggle against the enemy. In addition, the glorification of the warlike ethos is expressed in deliberately traditional language. Both the Cross of Sacrifice in the Tyne Cot Cemetery near Passchendaele—designed by Reginald Blomfield—and the Great Yarmouth memorial visibly embody the association of sacrifice and war, symbolized by a Christian cross and a bronze sword respectively. Likewise, the tower of the Douaumont Ossuary in Verdun, where the remains of more than 130,000 unidentified soldiers were buried, is shaped like an imposing artillery shell; with the four crosses on its walls, it dominates the field of battle (Dogliani 1986: 563). The form of the Menin Gate is no less apt. Although it was not meant to glorify war, it was nevertheless designed as a triumphal arch featuring not only the names of the war dead and a sarcophagus but also a massive lion, which stands, in Blomfield's words (in Stamp 2007: 103), for 'the latent strength and heroism of our *race*' (emphasis added).[53] The emphasis is on historical continuity with the old military tradition rather than merely on human tragedy. In these war memorials there is no hiding of the violent nature of the soldiers' duty, as they are often portrayed in defensive or offensive posture with their rifles, machine guns, and grenades. Furthermore, captured enemy artillery pieces are often located next to monuments.[54]

[53] The lion of the Menin Gate was probably inspired by the Lion's Mound on the battlefield of Waterloo.

[54] The practice of using enemy weaponry for constructing memorials will gradually decline after the Second World War. For an exception, see, for example, the pacifist Second World War memorial made

In spite of the probably painful feelings of grief and mourning experienced on a private level, this type of commemoration showed that soldier deaths were considered important but not problematic. Although bellicose war memorials are far from the majority among monuments dedicated to the First World War, the fact of showing soldiers with weapons in their hands seems, nevertheless, to imply continuity with earlier epochs and to suggest that the war made sense in a traditional manner.[55] Even though the loss of ordinary soldiers was commemorated, the way their death was portrayed remained largely within the boundaries of the traditional approach. The narrative of heroism was not replaced with one of human suffering; the latter was included in the former.

In all combatant countries, commemoration was still a political statement largely meant to dignify the nation's war effort and the nobility of the warrior ethos. First World War memorials are often monuments to battle and combat. As Mosse (1990: 6) explains: 'The function of consolation was performed on a public as well as on a private level, but in remembrance of the glory rather than the horror of war, its purposefulness rather than its tragedy.' Although French memorials are called *monuments aux morts* (monuments to the dead) rather than war memorials, the stress on death is nevertheless accompanied by glorification of the war effort.[56] Indeed, a typical theme of French monuments is a soldier delivering a bayonet thrust.

The dedications carved on the memorials are revealing as well. Phrases such as 'to the glorious memory of', 'the victorious dead', 'the heroic dead', 'in proud remembrance of', 'in honour of', 'Pro Patria' (for the nation), 'Pro Rege' (for the King), 'for England', 'for France', 'for Italy', and so on, all emphasize the positive outcome of the soldier's sacrifice (Moriarty 1997: 137; Sokolowska-Paryż 2012: 8–9). Even the tomb of the Unknown Warrior in London was erected in order to celebrate the signing of the Versailles Treaty on 19 July 1919 and to testify to the formalization of victory and the beginning of the post-war international order. As Alex King (1998: 142–3) clarified: 'it provided an object for the parading soldiers to salute in honour of their comrades who had been

from unexploded bombs in front of the church of Medesano near the city of Parma, Italy. But here enemy weaponry is meant to emphasize the human cost of war (borne by civilians in this instance) rather than the glory of the nation.

[55] The traditional links can be totally explicit, as in the Paisley memorial where British soldiers surround a mounted medieval knight. On medievalism in post-First World War Britain and Germany, see Goebel (2006). Something similar can also be found at the Royal Welch Fusiliers Memorial in Wrexham, where a First World War soldier is backed by an eighteenth-century one.

[56] Elaine Scarry (1985) famously argued that discourses on war, in order to displace the brutal reality of combat, tend to relegate the physical, mortal body of soldiers to the background and replace it with a harangue about the state and the army as immortal creatures. In actual fact, as far as the First World War commemorative practices are concerned, soldiers' bodies do not seem to have been overlooked at all. Their death has actually become the focus of memorialization and is given a positive meaning.

killed', but it was also saluted by 'Allied commanders, Foch, Haig and Pershing amongst them'. As a result, human loss and tragedy were set in the contexts of military victory and triumph.

In the case of France, the political use of the Unknown Soldier is even more striking. The burial ceremony in Paris was exploited to represent the victory of 1918 as a 'closure for the trauma' of the Franco-Prussian War, as illustrated by Varley (2008: 231) in this detailed description:

> Before being taken to the Panthéon, the Unknown Soldier was left for one night at Place Denfert-Rochereau, the square in the fourteenth *arrondissement* of Paris named after the hero of the siege of Belfort and the site of a monument dedicated to the Franco-Prussian War. The following day, the cortège was joined by an urn containing the heart of Gambetta, whose memory remained synonymous with the republican patriotism of 1870–1871 and fidelity to Alsace-Lorraine. On either side of the grave under the Arc de Triomphe were plaques bearing inscriptions that made the links between the two wars explicit: the first read, '4 September 1870 proclamation of the Republic', while the second declared, '11 November 1918, the return of Alsace-Lorraine to France'.

Finally, it should be stressed that monuments and memorials pictured the war not in its real filthy nature, but rather as a clean, noble endeavour. Unlike actual war and literary works such as Erich Maria Remarque's *All Quiet on the Western Front*, Robert Graves's *Good-Bye to All That*, Richard Aldington's *Death of a Hero*, John Dos Passos's *Three Soldiers*, and Emilio Lussu's *Un anno sull'Altipiano*, First World War memorials and the sanctifying speeches that inaugurated them bear no trace of mud, rats, and fleas. As argued in Chapter 2, the construction of memory is about both remembering and forgetting. In order to represent the war as a just and noble conflict, some of its dirty sides were kept out of memorial design and commemorative practices. Thus, the violent nature of war was not masked, but its less honourable aspects were utterly omitted.

In short, while war memorials were certainly meant to offer places for mourning and to preserve some material recollection of the dead, especially in Britain and in the Dominions, where they became a substitute for bodies and graves (Damousi 1999), their particular shapes and styles had traditional symbolic and political functions. Despite touching the chords of mass bereavement, they extended to common soldiers the old conventional view of the nobility of arms. In doing so, they did not subvert the time-honoured approach to commemoration; there was no rejection of a long-established language, but rather an extension to human beings who had previously been excluded from its domain. In other words, the individual was not yet elevated above the cause. The common soldier was recognized as a significant participant in war whose death was understood as a meaningful, justified sacrifice, not as unforgivable loss.

Two different processes were simultaneously at play. By commemorating the average soldier, the monuments marked a departure from the early modern period. By celebrating death within the traditional framework of the superior, noble cause of the war, however, they showed no break with the past. First World War memorialization, with its peculiar blend of political glorification of war and individualization of death, should not be interpreted as the beginning of a new practice. Instead, it marked a further step in a process that had started in the nineteenth century, a movement in which the traditional policy of glorifying war was not replaced but was combined with new attitudes towards soldier deaths.

The Natural Bodies of Soldiers in the First World War

Despite important changes in the disposal and commemoration of the war dead, soldiers' lives on the battlefield still 'counted for nothing' (Ellis 1986: 179). Not only were soldiers freely expended in the American Civil War and in the other major European conflicts of the period, but killing and sacrificing as many enemy combatants as possible became one of the central tactics of the First World War.

Battles lasted for several months rather than for hours and days, as they used to, and millions of people died on the various fronts of Europe and the Middle East. The human price was shocking: the war dead numbered nearly nine million and the wounded more than twenty million (Kramer 2007, 251). In Britain, more than 700,000 soldiers died during the war. The hundreds of thousands who were wasted became known as the Lost Generation. The French dead were approximately 1,300,000: 27 per cent of all men between the ages of 18 and 27 (Audoin-Rouzeau 1996: 221). Germany lost approximately 2,000,000 soldiers and Italy 651,000.[57]

[57] It goes without saying that these and other casualty figures are controversial and are likely to remain so in the future. The numbers mentioned in previous sections, in this chapter, and further on in this study should be considered approximate.

Understandably, the length, the intensity, and the number of casualties in the First World War redefined what western societies meant by war. The exponential increase in the killing power provided by machine guns, artillery, mines, and poison gas turned the First World War into the first real 'war of machines' (Neiberg 2006: 159). Soldiers had the sacrificial role of being killed by a vast arsenal of old and new weaponry. Not coincidentally, in the war the soldiers who perished of battle wounds outnumbered those who died of diseases acquired when in the army (Browning 2002: 135). Until the Franco-Prussian War, on average, less than 20 per cent of casualties were caused by enemy action, but in the First World War the proportion was dramatically reversed.[58]

Neither side could have imagined the death tolls or the length of the conflict; it was the prevalent belief that hostilities would be all over before the first Christmas. Europe's powers were positive that victory lay in the size of their armed forces and in the speed at which they could be deployed at the outbreak of the war. At that time, the main fear among volunteers all over the world was not of a long war, but that the conflict would end before they could join in. Little did they know that their enthusiasm was soon to be buried in the mud of the trenches.

Generals were not aware that wars of movement and manoeuvre were over. Ten years before the outbreak of the First World War, the Russo-Japanese conflict (1904–5) erroneously validated the impression among European military commanders that firepower could be overcome by numbers and by the superior morale of the attacking forces. However great the enemy's firepower, however great the organization of the enemy forces, offensive was superior to defensive; that was the common, mistaken belief that guided European generals (Howard 1986). The experience of the long and highly destructive American Civil War was utterly dismissed as irrelevant to the glorious military tradition of the Old Continent.

The German war plan, in particular, envisaged encirclement of French troops through movement and speed.[59] The idea was to win a war of manoeuvre without primarily relying on bloody frontal attacks—a plan to some extent similar to the strategy employed by Moltke the Elder against Austria–Hungary in 1866 and France in 1870. Initially, by reaching the Channel coast in Belgium and then moving towards France, the Germans seemed to be succeeding in

[58] The United States is an exception. In the First World War, more American soldiers died of disease (63,000) than of combat wounds (53,402). See Congressional Research Services (2019: 2).

[59] The German war plan was a modification of the so-called Schlieffen plan elaborated in 1905. According to some scholars, the changes were responsible for the stalemate that Schlieffen had tried to avoid in the first place (Rothenberg 1986: 320).

their strategy. The French victory at the Battle of the Marne (6–12 September 1914), however, meant that the Germans could not reach Paris, as originally planned. The attempt by both sides to outflank the enemy failed and showed the limits of manoeuvre in a war dominated by defensive weapons. The human cost for both sides at this point of the conflict was horrendous: the death toll was about 500,000.

By this time, it was clear that the war would not be over before Christmas. With the end of the first battle of Ypres in November 1914, the last opening in the western front was closed. Encirclement had become impossible, and mobile warfare ceased until March 1918, when the Germans, by resuming a war of movement, tried to reach total victory but found defeat after an initial and deceptive success.[60] Hence, for more than three long years, trench warfare became the predominant way of fighting in Europe.[61] Although in 1904–5 the Japanese army in Manchuria won by attacking the Russians, once the trenches had been dug during the First World War, the machine gun provided an accurate barrage of fire that rigidified the front. Being easy to hide and protect during an artillery attack and taking up only two feet of the battlefield, machine guns were hard to locate and suppress. Not only could they take enormous toll of advancing enemies but, by slowing down the attacking troops, they increased the soldiers' exposure to artillery fire (Dupuy 1984: 218). That is how the age of successful close-locked infantry attacks came to an end.

Life in the Trenches

Quite correctly, the typical image of the First World War is of trench warfare. Although many soldiers were killed in submarine and aerial confrontations, more men died under intense fire and shelling by artillery in a war of entrenchments. Trench systems were created facing each other and divided by so-called No Man's Land. Miles of trenches were dug on the western front: they stretched for some 470 miles, running without break from the Swiss border to the North Sea in Belgium, though most of the fighting was located west of Verdun, the major fortification on the western front.[62] The southern front, mainly

[60] The final stage of the war has been described as a combination of the mobility of the early phase with the firepower of trench warfare.

[61] One of the few examples of a successful breakthrough occurred on the Eastern Front, in the German-Austrian victory against the Russians during the Gorlice-Tarnów campaign in 1915.

[62] Conversely, the front line in the southern part of the western front did not see any large-scale operations throughout the war, even in the German area near Switzerland occupied by the French. In these sectors, trench life was characterized by a 'live and let live' attitude.

between Italian and Austrian–Hungarian forces, stretched for about 400 miles from Switzerland to the Adriatic Sea, but, including the ups and downs of the mountains, it stretched for 930 miles. With no flanks, the two trench systems did not offer any opportunity of enveloping the enemy forces; bloody frontal attacks were then regarded as the only possible military option. They were attempted, but with little or no success. The result was a bloody stalemate in which the infantry bore the brunt of the fighting.

The First World War will be remembered on account of the grotesque infantry assaults, which succeeded in acquiring few, if any, miles of battlefront at the price of hundreds of thousands of dead. However, most of the life spent in the trenches was not about combat, raiding, and patrolling but, on the contrary, about preparing and waiting for engagement. Digging, filling, and carrying sandbags were the most recurrent activities for a soldier during the First World War. Trench life has been accurately described as prolonged, vigilant inaction and combat, as one British soldier put it, like a 'mere incident between one set of billets and another' (in Winter 1978: 223).[63]

Because of the squalid living conditions, trench life was miserable. Rats, lice, fleas, and even rotten bodies were a constant presence. The worst enemies for soldiers were cold, wet, and mud. Staying warm and dry was one of the main challenges during the conflict. There were no winter holidays, and soldiers had to endure all the hardships inflicted by the weather. Rain turned the trenches into mud canals, and cold literally froze men to death in all armies. Conditions on the Alpine front in Italy were particularly harsh for soldiers, who were threatened not only by cold and snow but also by avalanches. Literary critic Paul Fussell (2013: 39) quite rightly described the trenches of the First World War as a 'troglodyte world'. Not surprisingly, such a life left painful scars and proved to be a lasting memory for those who survived.

Mass Death in Battle

Although no two battles were identical, First World War fighting was typically organized in two main stages: pre-battle artillery bombardment, which could last for hours or even for some days, and then repeated charges against the enemy lines.[64] After the initial artillery bombardment, which was meant to

[63] There were three main types of trenches: front, support, and reserve. Soldiers normally spent a third of their time in each.

[64] In theory, there should have been a third stage in battle: after the infantry had penetrated and destroyed any barbed wire, the cavalry were supposed to come through and inflict the final blow.

neutralize enemy fire, special units were sent towards the enemy trenches in order to destroy barbed wire with cutters. Soldiers were then ordered to fix bayonets and advance into the open field in open formation with no cover. In order to encourage them before an attack, troops were primed with nationalistic appeals and reminded about courts martial, but their fighting spirit was also boosted through the issuing of rum, cognac, wine, and chocolate. 'Rum and blood'—these were the prevalent smells in the air during an assault (Holmes 2005: 237).

Because massed infantry attacks were always preceded by long artillery barrages, surprise was virtually impossible. Thus, once the soldiers had left the relative security of the trenches, their bodies were fully exposed to lethal enemy fire. This was the greatest trial of battle in the First World War, when the men's vulnerability was total and their martyrdom almost inevitable.

The attack was then carried out with rifles, grenades, small mortars, and, after their introduction in 1915, flame-throwers. The tactic was aptly summarized in Pétain's slogan 'artillery conquers, infantry occupies' (Gudmundsson 1993: 43; Grotelueschen 2007,:30). The British were explicit in endorsing both the tactic and its deadly implications. In their *Infantry Training 1914* we can read that the 'main essential to success in battle is to close with the enemy *cost what it may* ... The object of infantry in attack is therefore to get to close quarters as quickly as possible' (in Holmes 2005: 232; emphasis added). However, magazine rifles, quick fire, and heavy artillery gave the defence an advantage unprecedented in the history of western warfare. The paradox about trenches is that they were designed to protect combatants, but, once they had been built, the only way to achieve breakthrough was by using massive attacks that would increase casualties enormously. As John Keegan (1999b: 293) pointed out, this condition was perfectly understood by generals: 'The simple truth of 1914–18 trench warfare is that the massing of large numbers of soldiers unprotected by anything but cloth uniforms, however they were trained, however equipped, against large masses of other soldiers, protected by earthworks and barbed wire and provided with rapid-fire weapons, was bound to result in very heavy casualties.'

Since generals believed that victory was a matter of numbers as battles had been won for hundreds of years, they increased the quantity of men deployed on the battlefield and the amount of guns and shells. Yet, as John Ellis (1986:

However, the presence of shell-torn ground and inefficiency in barbed-wire cutting, combined with the continuous barrage provided by machine guns and artillery, meant that cavalry were of little help on the western front.

179) specified in his classic study on the machine gun, they were disastrously wrong:

> Time and time again they threw their men forward, confident that this time a little more preparation, a few more men, and an extra dash of sheer courage would suffice to break the enemy's will to resist. They never realised that they were not fighting his 'will', but his machine guns. And they were implacable and unshakeable. Morale was an irrelevancy to them; all they needed was enough water and bullets. The man counted for nothing. The machine had taken over.

Although—or perhaps because—men 'counted for nothing', the main policy of the armed forces on either side was to cause more deaths to the enemy than they themselves would sustain. Killing as many enemy combatants as possible became one of the main tactics to achieve victory.[65] For example, the offensive of Verdun, from February to December 1916, was planned by the German military command as an attempt, in the brutal words of General Erich von Falkenhayn, to 'bleed' to death the French army (Philpott 2014: 212). Thus, the main goal here was not to conquer and occupy territory, but to kill and disable as many men as possible in order to inflict a fatal blow on the morale of the French army (Mombauer 2001: 210–13). The target was the Allied will to fight; the instrument was the physical destruction of its men, regardless of the tactical advantages acquired in the process. The outcome was horrific for those who fought that battle, and the figures were devastating: more than 700,000 were killed, missing, or wounded during a 299-day battle in which 24 million shells were fired by both sides.

Falkenhayn's so-called blood-mill approach was clearly meant to increase casualties in the expectation that mass slaughter would convince the French and British political authorities to end the war. He was proved wrong. Although the French army was ripe for defeat, Britain and France were prepared to bear even greater hardships, largely at the expense of their men. While most European powers had introduced conscription after 1871 (French 2005: 82), in January 1916 Secretary of State for War Kitchener started it in Britain for the first time in its history and called to arms hundreds of thousands of new

[65] The second central strategy, and according to some historians the victorious one (Osborne 2004; Ritschl 2005), was starvation of German society through an economic blockade imposed by the Royal Navy. William Philpott (2014: p. xvi), by contrast, suggests that the war 'ended decisively on the battlefield'.

men who would share the danger, the misery, and the fate of their willing predecessors.[66]

Far from being weakened in their morale and thanks to their new manpower, the members of the Entente devised the gigantic Somme offensive as a way to deflect the German army from Verdun. For a full week British bombardment prepared the Somme attack, which took place on 1 July 1916. In seven days, British guns fired 1,502,652 shells into the enemy lines (Middlebrook 1984: 105). The massive bombardment was intended to kill as many enemies as possible, to obliterate enemy positions and their barbed wire, and, in doing so, to enable the infantry to attack successfully. British hurricane bombardments, however, came to nothing; many shells and mortar bombs did not explode, and a shortage of heavy guns failed to blow away the barbed wire, which was untouched at many points of the German line. Moreover, holding the higher ground and with the region being well drained, the German infantrymen were able to build their deep dugouts thirty or even forty feet below ground and thus, during the seven-day bombardment, to take refuge in their relatively safe holes. Having failed to realize the depth and strength of German underground fortifications, British massed assaults met with tragic consequences (Philpott 2009: 191–2).[67] Contrary to all expectations, the Somme offensive was anything but a 'walkover'.

Having failed to destroy the enemy line, thousands of infantrymen marched to their death; the assaults *en masse* of eleven divisions encountered the fire of supposedly dead German machine gunners, who quite literally tore to pieces approximately twenty thousand British men and left another forty thousand wounded on the first day of battle. Wave after wave, the British infantrymen were cut down by a wall of German shells and bullets. That day, known among the troops as the 'Great Fuck-Up' (Fussell 2013: 13), was the bloodiest day in the military history of Britain: for 'every yard of the sixteen-mile front from Gommecourt to Montauban there were two British casualties' (Middlebrook 1984: 264).

Although an accurate picture of the butchery of the Somme is beyond description, graphic and vivid accounts of the carnage and terrible suffering endured by men were recorded by participants and diarists: machine-gunned

[66] Thanks to advances in farming techniques, not only could men be supported more effectively than in the past but there was also a decrease in the number of people needed on the land. Food would be one of the most important winning weapons of the First World War.

[67] One of the main reasons for the failure was the use of shrapnel bombs rather than explosive shells with a higher degree of penetration into enemy trenches. Shrapnel shells were able neither to cut barbed wire nor to destroy trenches.

soldiers, mangled bodies littering the battlefield, men disembowelled and blown apart, ragged amputations, heads taken off, splattering of human flesh everywhere, soldiers entombed by artillery shells, entire battalions wiped out in a few minutes' combat.[68] The confusion on the battlefield was so appalling that some British battalions, such as the Glosters and the Camerons, charged each other (Winter 1978: 179).[69] Eighteen years later, thinking about the first day of the Somme, the English poet Edmund Blunden wrote that 'Neither race had won, nor could win, the War. The War had won, and would go on winning' (in Brearton 2003: 50).

Regardless of the confusion and the horrendous German fire, the British infantry kept attacking. In spite of the huge impact of the Somme offensive on the British army, the carnage did not wreck British morale. After the first day of the Somme catastrophe, there was never a repeat of such a massive assault, but the fighting continued until November, when the battle ended with a British gain of six miles.[70] In the 141-day battle, soldiers on both sides of the front persevered in the face of enemy fire, and casualties inevitably mounted. British and German casualties totalled around 800,000.

The Entente's resolve did not diminish after the bloodbaths of Verdun and the Somme, because both France and Britain were in a position to replace the dead with fresh soldiers. Despite the heavy casualties, they had plenty of men available. Other attempts at breakthrough and major offensives were launched in battles such as Neuve Chapelle, Loos, Passchendaele, and Cambrai,[71] but it had become the main strategy to wear down the enemy's forces, because victory, it was believed, could be achieved only if one was prepared to sacrifice limitless numbers of men. As a prescient British military commentator put it in 1905, the 'chances of victory ... turn entirely on the spirit of self-sacrifice of those who must be offered up to gain opportunity for the remainder' (Maude 1905: p. x).

The return to mobility on the western front led to a further increase in conflict lethality. Thus, the tactical innovations in spring 1918 that allowed the Germans to conquer more territory than any army on the western front since

[68] See Keegan (2004: 242–69).

[69] So-called friendly fire killed thousands of men during the First World War.

[70] It should be borne in mind that the attack on the Somme relieved the French forces from defeat at Verdun where in June 1916 they seemed to be in a state of collapse. Of course, one might argue that such a success was not worth the appalling human losses of the Somme offensive. Reflecting on the deadlock on the western front, Winston Churchill (2005: 291) famously remarked that 'victory was to be bought so dear as to be almost indistinguishable from defeat'. See Hart (2008) and Philpott (2009) for reappraisals of the Somme.

[71] Some of these attempts, such as the one at Cambrai, were successful, but could not be exploited owing to lack of reserves.

the beginning of the war caused horrendous casualties. Between March and July 1918, official figures number the German dead, wounded, and missing at 915,000 (Smith et al. 2003: 151).

In the final phase of the war, the American army fought no differently from European troops. Although American soldiers had been employed in vast numbers only since the summer of 1918, they fought fiercely and with scant regard for human cost (Ferguson 1998: 312). In just six months on the front, 53,402 American soldiers were killed (Congressional Research Services 2019: 2).

Democratization of Death on the Battlefield

In the First World War, the traditional sharp division between officers and troops was displaced by a further distinction between staff and line officers. While staff officers (and of course the so-called *château generals*) were safe in their locations well behind the front, line officers shared the dangers and hardships of their infantrymen. Indeed, losses among line officers could be shattering. In the French infantry, casualties among junior and non-commissioned officers averaged 25 per cent against an 18 per cent death rate for the enlisted men (Smith et al. 2003: 69, 96). The casualty rate among line officers even reached 75 per cent, as occurred in four British battalions in 1914. On the first day of the Somme, 75 percent of officers were killed or wounded (Middlebrook 1984: 263). Moreover, some of the casualties belonged to the most important European and American families. To quote a few notable examples: Raymond Asquith, officer of the Grenadier Guards and son of the British prime minister Herbert Asquith, died at the Somme in September 1916; Rudyard Kipling's son John was killed at the age of 18 during the Battle of Loos in September 1915;[72] in December 1915, Émile Durkheim's son André died of war wounds in a Bulgarian hospital (Coser 1977: 149); Quentin Roosevelt, the youngest son of former president Theodore Roosevelt, was killed in aerial combat over France on 14 July 1918 (Trout 2010: 222).

As these examples illustrate, the sons of the social and intellectual upper class were not only involved in the war, but also paid a high price in terms of human lives. The scale of death for officers meant that a considerable number of them were not selected just from the 'gentlemen's class'. The increasing need

[72] Because of poor eyesight, John Kipling succeeded in obtaining a commission in the Irish Guards only thanks to his father's influence.

for soldiers, due to the lethality of the First World War, opened the senior ranks to the middle classes and encouraged the democratization of both the army and death on the battlefield.

The Cheapness of Soldiers' Lives in the First World War

The military tactics employed during the conflict showed that generals and statesmen had little or no regard for their men. Verdun, the Somme, Loos, Gallipoli, the eleven offensives on the Isonzo river, Caporetto, and many other battles were epic clashes in which soldiers did more than their duty and demonstrated an impressive devotion to their countries. They fought hard; they sustained grievous casualties in the attempt to push into the enemy lines; and they were usually bloodily repulsed. The degree of their devotion was second only to the degree of indifference shown by their commanders, who displayed no mercy in the way they employed their men. Under the leadership of generals such as Charles 'the Butcher' Mangin, Douglas Haig, Luigi Cadorna, and many others, the lives of soldiers were held as cheap, and pointless death often occurred. What Keegan (1999b: 289) wrote about General Haig seems to apply to many of his foreign colleagues: in his 'public manner and private diaries no concern for human suffering was or is discernible'. In the British army before the spring of 1916, some commanders even forbade their soldiers to use steel helmets (introduced at the end of 1915), because they were afraid that would reduce toughness on the battlefield (Middlebrook 1984: 38). In fact, early in the conflict, the British soldier was sent to war 'dressed like a gamekeeper in a soft cap' (Bourne 2005: 133).[73]

Trained volunteers and drafts of less-trained conscripts were alike treated like chess pieces that could be moved and sacrificed at whim, not as human beings with specific, individual rights. Whether professionals or conscripts, soldiers remained faceless and interchangeable, and their employment was characterized by a total lack of regard for their personal safety.

Despite remarkable improvements in the legal protection of the wounded, evacuation time (motorized ambulances), and medical knowledge, and notwithstanding stalwart work by the different national sections of the Red Cross, the disregard for soldiers in combat was almost the same as that for the injured. While medical treatment today is given first to the critically injured,

[73] Likewise, the French soldiers of summer 1914, dressed in red trousers and blue overcoats, were easy targets for the Germans. As Alastair Horne (1981: 25) put it: 'Never have machine-gunners had such a heyday. The French stubble-fields became transformed into gay carpets of red and blue.'

during the First World War it was standard policy to pick the less seriously wounded first and take them to casualty clearing stations. The rationale was to provide medical assistance to men who could be put back on the killing fields. 'In the same spirit', writes Denis Winter (1978: 196), 'priority of movement in the trenches went first to ammunition, second to reinforcements', and only then to the injured. Because truces to evacuate the wounded were seldom tolerated by commanders, thousands of combatants died owing to lack of immediate medical treatment. Thus, the priority was winning the war, not saving or protecting soldiers' lives. Humane considerations were still secondary to military necessity. Not surprisingly, facts and vivid memories reported by historians, commentators, and former combatants have exposed the cynical attitude of generals to the conditions of their men. Remembering the carnage of the Somme, for example, Sgt C. E. Linford said that it 'was pure bloody murder. Douglas Haig should have been hung, drawn and quartered for what he did ... The cream of British manhood was shattered in less than six hours' (in Middlebrook 1984: 316).

By today's standards, wantonly sacrificing human lives by launching men to almost certain death is unquestionably criminal behaviour. The main problem with such a view, however, is not only that it is largely divorced from the realities of war, but also that it lacks historical perspective. Generals' attitudes may appear to have been criminal, and even sadistic to some observers, but only in relation to our present standards, certainly not in comparison to how soldiers had been employed before the First World War.[74] As shown in Chapter 2, commanders in early modern Europe inflicted massive sufferings on their soldiers to achieve military advantages. Achieving victory on the battlefield was the ultimate goal of warfare, and any sacrifice could be potentially justified.

The cynical treatment of infantrymen in the First World War shows that attitudes towards the natural bodies of soldiers as *human material* had changed little from prior centuries. Some of the features of nineteenth-century warfare and First World War combat distinguished them from previous wars, particularly in the area of military technology, but the way western governments expended their men was still the same as before, with no real distinction between democratic and non-democratic regimes. It should also be emphasized that the type of military recruitment did not affect attitudes towards soldiers'

[74] Moreover, generalizations should be avoided and reservations allowed concerning military commanders' insensitivity towards their men's lives. For positive re-evaluations of General Haig, see Hart (2008) and Sheffield (2016).

lives.[75] Not only did most democracies waging the First World War use forms of conscription without being sensitive to casualties, as testified by the suicidal tactics employed by democracies on the western front, but the case of Britain, a democracy that entered the war with a volunteer army and introduced conscription in 1916 with no apparent change in aversion to its soldier toll, casts serious doubts on the commonsensical idea that the political regime in itself or the type of military recruitment actually affects attitudes towards soldier deaths. Even though the men's conditions had improved significantly in comparison to the past, when armies were decimated by starvation and pestilence, the value of a soldier's life was still restricted to his fighting capacities.

Conclusion

One of the main paradoxes of the American Civil War and the First World War lies in the fact that living soldiers were treated with no or little care. In death, instead, their corpses were handled with great consideration. As Denis Winter (1978: 260) wrote in relation to the First World War: 'It seemed that no effort was too much to care for the bodies of men who had been handled with so little regard while they had been alive.' Moreover, after the war had ended, such care for the dead also prevailed over other important social needs of the living, such as housing, employment, and public order (Capdevila and Voldman 2006: 27).

The almost total disregard for soldiers' lives shown by western governments means that—in terms of men's employment on the battlefield—the period from the American Civil War to the First World War had more in common with the age that preceded it than with the one that was to follow. On the other hand, the creation of military cemeteries and the memorialization of common soldiers marked a profound fracture with the past. For the first time in the modern period, ordinary soldiers were both lamented and celebrated, and their death took on public and social relevance. In other words, though combatants still died 'as cattle', to use Wilfred Owen's words,[76] they were granted a truly social body.

The fact that the narrative ploy of sacrificial death was devised by states as a symbolic policy with a view to appropriating the immolation of their soldiers (Bogacz 1986: 651) does not imply that it had no long-term effects. Notwithstanding attempts by governing elites to reap political capital from

[75] On the supposed relation between conscription and casualty sensitivity, see the opposing views of Moskos (2001) and Smith (2005).

[76] The words are quoted from the opening line of Owen's 'Anthem for Doomed Youth'.

the construction of military cemeteries and war memorials and regardless of the rhetorical nature of public ceremonies, through these material and cultural devices western states symbolically confirmed the emerging social body that ordinary soldiers had previously lacked. With the American Civil War, the Franco-Prussian War, and especially the First World War, soldiers ceased to be anonymous collections of men and became individuals with a story to be told. As shown in Chapter 4, this publicly recognized sacrifice contributed to a number of unintentional and unforeseen results. Among these, it helped to turn soldiers from mute sacrificial objects into individuals with personal rights and empowered them against the idea that wholesale death in war is a natural and therefore acceptable occurrence.

4
Irrecoverable Death

As we have seen, the process of sacralization and individualization of the fallen illustrated in Chapter 3 did not concern western combatants when they were alive. However, in the decades following the First World War that process involved not merely the (dead) social body of soldiers but also their natural, living bodies when deployed on the battlefield in order to fight. In this period, the sacralization of the individual soldier is shown in all three areas under study: disposal of corpses, commemoration, and soldiers' employment in war. In this age there is a convergence of the three practices, which had departed from each other in the epoch of transition. These changes in the meaning attached to soldiers' lives and deaths, far from being merely ritualistic and cosmetic, have profoundly affected how western states wage war.[1] For, as the cold war proceeded, western societies became gradually more sensitive to friendly casualties, and their armed forces adopted a set of practices meant to reduce their human losses.

Sparing Soldiers' Lives: The Targeting of Civilians in the Second World War

When the First World War came to an end, with nearly nine million dead and more than twenty million wounded, a widespread revulsion against war and weaponry emerged in countries such as France, Britain, and the United States, in which the prospect of another military catastrophe seemed utterly unthinkable. Yet, twenty-one years later, the Second World War broke out. Marshal Ferdinand Foch's prediction after the Peace of Versailles—'This is not peace; it is an Armistice for twenty years' (in Kissinger 1994: 250)—proved tragically correct.

[1] Although the empirical evidence—as elsewhere in this book—regards a variety of states, this chapter, especially the sections on warfare, will focus primarily on the American experience inasmuch as, since the Second World War and especially after the end of the cold war, the United States has become the most active military power in the world.

Life, Death, and the Western Way of War. Lorenzo Zambernardi, Oxford University Press.
 DOI: 10.1093/oso/9780192858245.003.0005

Overall, casualties in the Second World War were much higher than in the previous conflict, as the dead worldwide numbered between fifty-five and sixty million. However, though weaponry was greatly more lethal and the war deaths increased as a whole, military casualty rates did not reach the levels of the First World War, with the sizable exceptions of the United States and combat on the Eastern Front (Clodfelter 2008: 2).[2] Notably, unlike the First World War, casualty figures among civilians in the Second World War exceeded those of combatants. While warfare in the First World War usually took place in pastoral landscapes and tended to be a strictly military affair between armies—with important exceptions such as Louvain, Ypres, and Albert—urban centres and their residents were heavily and wantonly targeted during the second world conflict. Large-scale indiscriminate violence against civilians, unleashed through bombing, rocket attacks, and ground operations, became the norm. Even the strategic bombings carried out by two liberal democracies like the United States and Britain were deliberately aimed not only at the enemy's forces and military installations, but primarily at the civilian population and infrastructure. As a result, the traditional distinctions between combatants and civilians and between the front and the rear zone became fundamentally blurred.

The massive civilian targeting of the Second World War is generally explained as the logical consequence of the combination of total war and innovation in military technology. First, developments in air power extended the battle zone and brought the civilian population into the firing line. While civilian targeting in the past could occur only either when an enemy territory was occupied or at the end of a successful siege, progress in military technology led to unprecedented ways of killing at a distance. Not coincidentally, before the war, almost every major western country had had its own proponent of strategic bombing. Clément Ader in France, Giulio Douhet in Italy, William 'Billy' Mitchell in the United States, and Hugh Trenchard in Britain all argued in favour of bombers flying and striking beyond enemy lines. Secondly, because states had mobilized most of their material and human resources, the enemy population as a whole became the natural object of violence. Indeed, targeting the enemy population became a military corollary in total warfare just because civilians played a significant role in the war effort.

As regards the strategic bombings implemented by liberal democracies, three additional reasons are put forward to explain the controversial decision

[2] In the Second World War—unlike Germany, the Soviet Union, and the United States—European countries such as France, Italy, and the UK incurred fewer losses in terms of military personnel than during the First World War.

directly and wantonly to target the enemy population. First, the targeting of civilians was seen as a legitimate retaliatory act against German bombings. Secondly, early in the war, neither the Americans nor the British had enough troops to mount an invasion of continental Europe. In fact, when Britain began its strategic bombing campaign, there was no offensive alternative. Thirdly, bombing cities was a more brutal variation on the typical strategy of naval powers—that is, economic blockade: in order to win, one needed to defeat both the war-willingness of the enemy society (the British approach of breaking the enemy's will) and its war machine (the American approach of destroying the enemy's means of production) rather than crush its combatants on the battlefield.[3]

This threefold explanation may account for the British and American decision heavily to target enemy civilians and infrastructure. However, there is one more reason, which is usually given as a margin note but which seems to have played a significant role in determining the aerial strategy and the more general substitution of technology for manpower: by bombing from on high, the appalling friendly casualties of the First World War could be avoided. In other words, the use of air power, either in support of, or independently of, ground forces, prevented the trench stalemate and the horrendous casualties of the first conflict. In the Second World War, the heavy reliance on long-range bombing and the relatively cautious use of infantry, which was deployed after massive artillery and air campaigns, seem to have resulted from a deliberate attempt to avoid the deadly frontal assaults of the First WorldWar (Clodfelter 2010a).

The goal of sparing soldiers' lives was evident in a variety of situations. For instance, such a preoccupation very probably inspired the Americans' rejection of Churchill's request to speed up the advance of US troops in order to take Berlin and Prague ahead of the Soviets. Concerning the liberation of Prague, General Marshal declared that 'I would be loath to hazard American lives for purely political purposes' (in Matloff 1994: 534). It is noteworthy that such a cautious approach was a complete contrast to the Soviets' attitude. In the final phase of the war (16 April–8 May 1945), the Red Army suffered more than 300,000 casualties through using military tactics that did not aim to minimize its human losses (Erickson 1983: 621–2). Because the Soviets were not shy of

[3] British strategic bombing aimed to break the will of the people; American bombing was primarily employed to destroy the enemy's productive capacity, such as power plants, dams, infrastructure, and factories. However, the US Air Force also took part in area bombing of German cities such as Hamburg, Dresden, and others. On the two different strategic bombing doctrines and results, see Clodfelter (2010b).

expending their combatants in their relentless advance, they did not opt for strategic bombing, but rather used air power in close support of aggressive land operations (Overy 1987: 58).

Likewise, the way the Americans countered the Japanese willingness to fight to the death in the Pacific area was not a similar choice of self-sacrifice, but the decision to drop two atomic bombs on the cities of Hiroshima and Nagasaki. For, after suffering heavy losses in capturing the islands of Iwo Jima and Okinawa in early 1945, the Americans realized that an invasion of Japan would result in too many casualties.[4] The consequences were catastrophic for Japanese civilians. Low estimates deriving from the assessments of the 1940s suggest that the explosions immediately killed 110,000 people (around 70,000 dead at Hiroshima and 40,000 at Nagasaki); high estimates from the 1977 re-estimation put the number at 210,000 (140,000 dead at Hiroshima and 70,000 dead at Nagasaki).[5]

That controversial decision is still a source of divisive historical and ethical debate. According to some, it was an unjustified criminal act against unarmed civilians, based on a mixture of revenge and racism (Dower 1986). For Paul Nitze (1989: 43), instead, it was the equivalent to an air raid with incendiary bombs. For Thomas Schelling (1966: 17–18), it was merely a new chapter in the long history of 'coercive violence'. Others still argued that the nuclear attacks not only diminished American military casualties, but paradoxically saved thousands of Japanese civilians, since the alternative to the two nuclear attacks would had been an invasion preceded by heavy fire-bombing of Japanese cities (Hastings 2007: 504–40). Indeed, more people were killed in the air raid on Tokyo in March 1945 than in Nagasaki on 9 August (Neillands 2001: 378; Ralph 2006: 495).[6] Therefore, while it is not totally clear whether the nuclear attacks or a conventional bombing campaign followed by an invasion would have killed fewer Japanese civilians, what we know for sure is that

[4] On the Japanese will to fight to the death, it should be borne in mind that, at Saipan and Iwo Jima, Japanese prisoners surrendered in a ratio of one prisoner to every 120 dead. Allied soldiers, by contrast, surrendered in a ratio of one prisoner to every three dead (Buley 2008: 72). The traditional samurai ethics (*bushido*), which asserted the importance of an honourable death in war, seem to have played a significant role in convincing Japanese soldiers that dying was better than being captured. As Keegan and Holmes (1985, 50–1) aptly explained: 'Relatively few British and American units in the world wars defended a position literally to the last man and the last round: usually, once a certain proportion of casualties were sustained … the survivors fled or surrendered. For the Japanese, though, defence to the last man was routine, and attacks were pressed home, time and time again, in the face of certain death.'

[5] For a discussion of these figures, see Wellerstein (2020).

[6] An estimate of 100,000 Japanese people died during the bombing of Tokyo, when napalm and magnesium caused gigantic deadly storms of fire. However, it should be noted that Tokyo was a much more populous city (6,558,161) than Hiroshima (336,483) and Nagasaki (270,113). Data refer to the National Census of February 1944.

the decision to drop the nuclear bombs saved the lives of thousands of American soldiers, a significant concern for President Truman in the final months of the war (MacEachin 1998: 22). As General Curtis LeMay explained: 'if you don't destroy the Japanese industry, we're going to have to invade Japan. And how many Americans will be killed in an invasion of Japan?'(in Kozak 2009: 215). Nuclear weapons provided Washington with the solution to the dilemma of whether to reduce friendly casualties or to obtain a total Japanese defeat. In this sense, the use of nuclear weapons avoided an invasion of Japan that would have cost 'at a minimum one quarter of a million casualties' (MacEachin 1998: 33).

Moreover, the connection between the atom bomb and the possibility of saving American lives was also manifest in General Maxwell D. Taylor's reaction when he was informed by General Marshall about the existence of the nuclear weapon. Recalling an exchange with General Patton, Taylor remarked: 'What if we had such things to clear our way across Europe? Think of the thousands of our brave soldiers whose lives might have been spared' (in Lewis 2012: 65).

Although it would be an overstatement to argue that long-range bombing and nuclear attacks in the Second World War were early instances of what is now termed casualty aversion—for both the United States and some of its allies sacrificed large numbers of men to defeat the Axis powers—a newly emerging form of sensitivity towards military casualties seems to have played a significant role during the war. While bombing civilians in the Second World War, justified by various arguments, was certainly aimed at punishing and coercing the population, it was also intended to reduce friendly casualties. That is why, from the viewpoint of soldier expendability, the First and Second World Wars cannot be taken as a single historical episode running from 1914 to 1945. While most western governments in the First World War did not show any restraint in expending their men, the Second World War marked the dawn, or early infancy, of casualty sensitivity. As we shall see, the Second World War started a process of transformation that would ultimately lead to the marginalization of ground forces on the battlefield. On this point Adrian R. Lewis (2007: 37, 42), author of a study on the American culture of war, has stated that the 'shift toward greater use of fire-power' was the result of several variables, but 'the primary reason was to reduce casualties and save lives'. Although Lewis might have overstated the importance of friendly casualties—the Second World War cannot be labelled as the conflict in which casualty sensitivity fully emerged with all its effects on warfare—he appears to be perfectly right in seeing therein the first signs of a new way of fighting based on the removal of American and, more generally, western soldiers from the battlefield.

Second World War Memorials

After the Second World War, numerous memorials were erected all over the world: in capitals, cities, villages, and, as in the First World War, on major battlefields. Memorials raised to war heroes and political leaders are also to be found commemorating the Second World War, both in the United States and in Europe. In Westminster Abbey in London, for example, there is a memorial placed in 1965 in honour of Winston Churchill, Britain's leader for most of the war, and another statue of Churchill stands in Parliament Square. British generals such as Alexander and Montgomery, as well as foreign allies such as De Gaulle, Eisenhower, and Roosevelt, have their own statues and plaques.

Despite this rather traditional approach to memorialization, which can be found in all major belligerent countries, Second World War commemorative practices, like those in the aftermath of the First World War, placed the emphasis on the ordinary war dead. Indeed, in terms of memory, the two world wars might seem to be part of a single story whose language and representation did not vary all that much. In this connection, Jay Winter (1995: 8–9) rightly maintained that the key feature of commemoration in the period between the two wars was continuity, not change.[7] For example, the American military cemeteries at Anzio/Nettuno (Sicily-Rome American Cemetery) and in Florence/Impruneta testify to continuity in the commemoration of the two wars. Here nothing suggests desperation and suffering; through a variety of architectural devices (for example, flowers forming stars, a relief showing a fallen soldier in the Madonna's arms, a statue of an armed soldier, vertical white architecture, and so on), death was framed as a noble sacrifice for the nation and as a celebration of victory.

Although the actual caesura with past honouring practices would occur only in the decades following the Second World War, when celebration of the patriotic certainties of glory and sacrificial death finally came to an end, there are least three main differences between First and Second World War memorials. First, the Second World War did not witness the mushrooming of memorial constructions that had been so typical of the earlier conflict. In a study on British memorials (Boorman 1995: 1), almost 700 are listed for the Second

[7] Little wonder that this time the Soviet Union, too, erected hundreds of memorials celebrating victory in the Patriotic War (Ignatieff 1984; Merridale 1999). Some of the Russian monuments are small; others are gigantic, like the one commemorating the epic battle of Stalingrad on Mamayev Kurgan in Volgograd (known from 1925 to 1961 as Stalingrad) and that in Treptower Park in East Berlin. However, Soviet memorials are somewhat different from west European ones. In particular, they commemorate the cause of the Great Patriotic War with general reference to the fallen ('Here lie the people of Leningrad' states the memorial at Piskarevsky) but without an inventory of the dead. As Valentin Bogorov observed: 'the first Soviet war memorials … celebrate the prowess and the newly redefined image of the Soviet state' (in Kattago 2009: 157).

World War, a significant but tiny number if compared to those erected for the First. Even in the United States, the conflict was not greatly commemorated, and, when it was, traditional monuments were often replaced with living memorials.[8] Arlington's famous Iwo Jima Memorial, officially dedicated by President Eisenhower in 1954, is an exception rather than the rule. Moreover, it is significant that no national memorial was erected in honour of the fallen of Pearl Harbor until 1962. Likewise, the National World War II Memorial in Washington, located in the National Mall between the Lincoln Memorial and the Washington Monument, was not proposed until 1987. After a long and difficult process of authorization, it was opened to the public in 2004.[9]

In most cases, First World War memorials were rededicated and adjusted to include the new generation of war dead by adding further panels with the names of the newly fallen (Saunders 2001: 484; Dogliani 2006: 265). In the United States, Armistice Day (11 November) was renamed Veterans Day in order to combine the commemoration of both world conflicts.[10] Here the process of merging the memory of the two world conflicts was finally completed in 1958 when the Unknown Soldiers of the Second World War and Korea were buried beside the First World War Unknown.

Secondly, one of the peculiarities of Second World War memorials is their inclusion of the civilian population in the process of commemoration. Not only were the names of the members of the various national Resistance movements inscribed on new monuments and plaques in countries such as France and Italy,[11] but also included were those of the civilians killed during enemy operations. In Britain, for example, the names of about 66,000 civilian victims of enemy actions are recorded in a roll of honour placed in St George's Chapel in Westminster Abbey.[12] Several local memorials in Italy and France likewise bear the names of civilians. The fact that common people were massively

[8] Living memorials such as stadiums, parks, schools, and auditoriums had become more fashionable than traditional war monuments (Piehler 1995: 134–7).

[9] I will discuss the structure and content of the National World War II Memorial in Washington later in this chapter.

[10] Since 1954, 11 November has commemorated all the American war dead.

[11] In Piazza del Nettuno, on the wall of Palazzo D'Accursio at the centre of the Italian city of Bologna, there are pictures of the local partisans killed during the war of liberation. This shrine resulted from spontaneous initiative by the Bologna populace, who, as of 21April 1945, began to lay flowers and leave pictures of the local fallen. The location of the shrine was not coincidental; it was the site at which for months the Nazis and the Black Brigades had carried out summary shootings of partisans and antifascists. For one of the few studies that catalogued memorials dedicated to the Resistance in Italy during the cold war, see Galmozzi (1986). By no means coincidentally, the vast majority of these memorials are located in Northern Italy (with the notable exceptions of Naples and a few towns in Abruzzo), where most partisan warfare took place (Dogliani 2006: 266).

[12] It was the founder of the Imperial War Graves Commission, Sir Fabian Ware, who proposed the idea of commemorating civilians.

targeted in this conflict probably accounts for their inclusion in the process of commemoration.

Thirdly, abstract images increasingly replaced human figures (Winter 2006: 151), and, more importantly for the present research, the new inscriptions on a number of monuments conveyed slightly different values from those expressed in the aftermath of the First World War. The language of glorious death began to be replaced by a representation of death as still justified by the cause, but death was no longer portrayed as a beautiful sacrifice for the nation. To mention just one case in point: the inscriptions on the war memorial at Ledbury, Herefordshire, reveal that something had changed since the First World War. The earlier dedication, typical of many First World War memorials, proclaims the glory of dying for Britain in war: 'To the Glory of God and to the immortal memory of the gallant men from this town who gave their lives for their country in the Great European war 1914–1919. Greater love hath no man than this.' By contrast, the inscription dedicated to the fallen of the Second World War reads simply: 'Let us remember before God the men of this parish who gave their lives in the cause of freedom 1919–1945' (in Bushaway 2001: 502). Both captions justify death by referring to the noble causes for which the war was fought (the country and the defence of freedom, respectively), but the later one describes death as loss and no longer as glorious action. The different wordings seem to imply that dying in war had at least partly lost the glory that had previously been attached to it. This subtle shift can be found in many British Second World War memorials, where a soldier's death began to take on a tragic meaning in its own right, thus replacing the motif of sacrificial immolation.[13]

Such a change is also visible on the wall of the city of Volterra in Tuscany, Italy, where in Viale Vittorio Veneto there are seven war memorial plaques in remembrance of the (Second) Italian War of Independence (1859), the battle of Adua (1896), the First World War (1915–18), the Second World War (1940–5), the Liberation of the city (9 July 1944), the Italian Resistance (1943–5), and those killed in the Nazi camps (1943–5). While on the first three plaques (that is, the War of Independence, the battle of Adua, and the First World War) the language employed refers to the glorious sacrifice of the fallen soldiers with expressions such as 'holy wars', 'heroic sacrifice, 'martyrdom', 'greatness of the homeland', on the other plaques relating to conflicts after the First World War the language is merely descriptive (for example, 'from this door the "Allied"

[13] See also the Great Yarmouth War Memorial where the words 'king' and 'country' disappear in the inscription dedicated to the fallen of the Second World War. For further British examples, see the local memorials recorded in the project 'War Memorials' of The Coldstream & District Local History Society, https://www.coldstreamhistorysociety.co.uk/category/projects/war-memorials/.

soldiers entered Volterra') or refers only to the dead (for example, 'in memory of the fallen in the war 1940–1945'; 'those who fell in the resistance against Nazifascism').

It might sound like a riddle that the Second World War was much clearer than the First in terms of political responsibilities and more definitive in terms of its outcome (total victory with the invasion and occupation of the defeated enemies' territory), whereas war memorials in victorious countries were more about human loss than about the celebration of the cause. But, as already hinted, the key to the riddle may lie in the fact that, unlike the First World War, the Second World War blurred the distinction between combatants and non-combatants. And, yet, the circumstance that victory against the Axis powers was total and morally unquestionable ought surely to have fostered a note of heroic celebration in war commemorations, especially in the light of those who had perished for such a victory. A more likely reason for the paradox may be linked to the fact that the meaning of soldier deaths was in the process of change, with gradual modifications that would become more apparent in the following decades.

Post-Second World War Commemoration: From the Sacredness of Death to the Sacredness of Life

The changes in the nature and form of commemorative practices became more evident with the passing of time. After the Second World War, the theme of heroism in the symbolic representation of war does not completely disappear but is gradually marginalized. While the general themes of war memorials in early modern Europe were power and victory, and after the First World War glorification of the death of common soldiers, commemoration in the post-Second World War era has centred largely on death as an unforgivable loss. In Hélène Puiseux's words: 'In the course of two centuries, the representation of war has changed from a heroic depiction of conflict to a heroic depiction of loss, then to the loss of heroism and to the loss of any meaning' (in Capdevila and Voldman 2006: 15). As Reinhart Koselleck (2002: 322) noted, post-Second World War memorials are 'negative monuments', or, as Nathan Glazer (1996: 27) put it, they are 'mute monuments'. There is comparatively little depiction of power, victory, or the war cause; and, when there is some, the human cost of war is never forgotten. Post-Second World War memorials rarely highlight enemy-killing actions or, more generally, belligerence. In the words of Koselleck, they represent death in war 'only as a question and no longer as an answer, only as demanding meaning and no longer as establishing meaning. What remains is the identity of the dead with themselves; the capability of memorializing the dead eludes the formal language of political sensibility' (in Olsen 2014: 279).

In terms of visual representation, war as a glorious killing enterprise has been replaced by an image centring on pain and loss. Accordingly, not only have victims supplanted warlike heroes, but soldier deaths are no longer presented as a source of national salvation, and, as a result, western soldiers are portrayed as the victims of violence rather than as its active, noble perpetrators. Interestingly, this is also true for those Italian memorials dedicated to partisans, who are irrevocably portrayed as victims of violence (Isnenghi 2005: 326) rather than people using violence for a noble cause. Italian partisans look like martyrs without blood on their hands.

The downgrading of the war cause in favour of the fallen will come into focus if one looks at recent war memorials in European countries. In the British Armed Forces Memorial, located in Alrewas, Staffordshire, and officially dedicated in October 2007, the list of names carved on the walls includes not only soldiers killed in combat or during peacekeeping operations, but also those who perished in training and exercise—that is, without fighting. Moreover, in the two main sculptural groups that form the monument, no weapons or combat scenes are in sight. The group 'The Stretcher Bearers' is formed of a wounded man being carried by soldiers in the presence of a grieving family; the second one, 'The Gates', shows a fallen soldier helped by his comrades. War is portrayed merely as an event of pain and loss rather than a noble fight on behalf of the country. Here, there is no longer the belligerent religiosity typical of First World War memorials, as found, for example, in the military ossuary of Pasubio, Italy, where in the chapel frescoes a sword-bearing Archangel Michael presides over Italian soldiers (namely, the heavenly militia of God's angels) fighting against the Austro-Hungarian forces (namely, Lucifer).

The increasing consideration for the dead at the expense of the nation and the war cause is even clearer in the changing history of war commemoration in Germany, epitomized by the Neue Wache located in Berlin. This monument was originally erected between 1816 and 1818 as the 'new guard house' (*Neue Wache*) for the troops of the Crown Prince of Prussia, but its dedication changed four times in the course of the twentieth century. In 1931, it was dedicated to the fallen of the First World War; during the Nazi regime it became a monument to the 'undefeated heroes'; in the 1950s, the government of the German Democratic Republic rededicated it 'to the victims of militarism and fascism'; in 1993, the building finally became a memento to 'the victims of war and tyranny' (*Opfer von Krieg und Gewaltherrschaft*) with an enlarged version of a statue by Käthe Kollwitz that shows a mother cradling her dead son (Lüdtke 1997: 149–50). Although the political messages embodied in the Neue Wache have drastically changed over time (that is, German nationalism, Nazi revanchism, communist propaganda against capitalist imperialism, and finally pacifism), the only feature that has stayed the same is the centrality of the war dead, with the meaningful disappearance of the war cause and condemnation of all violence as the final message. In this monument, only human loss is now commemorated; the category of 'victims of war' allows no distinction either between soldiers and civilians or between friends and enemies: the 'common theme linking all of the dead … [is] … one of irrecoverable death' (Kattago 2009: 157). The final shape of today's Neue Wache seems, therefore, to be not a mere product of conservative identity politics meant to rehabilitate the German past, as Koselleck famously argued,[14] but the exemplar of a general change in the meaning attributed to death in war.

Interestingly, in German Second World War cemeteries located on foreign soil, such as the one at the Futa Pass between Bologna and Florence, Italy, one reads the same words as used in the Neue Wache: 'Opfer von Krieg und Gewaltherrschaft'.[15] Moreover, this cemetery contains the corpses not only of German soldiers who invaded Italy, but of members of the 16th SS Panzergrenadier Division 'Reichsführer-SS' who fought along the German defensive line on the Apennines, where many brutal massacres and slaughters of civilians took place (Dogliani 2009). Although this and other German cemeteries represent an act of reconciliation between two countries that have become friends since the tragedy of the Second World War, here again the traditional distinction between victors and vanquished, between the victims of war and its perpetrators, seems no longer relevant to the present.

All these monuments and cemeteries deny the positive function of killing and dying and invert the priority of nation over individual that was typical of First World War memorials. They show how the myth of glorious death in war, once regarded as the supreme and most noble sacrifice, has been replaced by

[14] Koselleck criticized the choice of the word *Opfer* (whose meaning is both victim and sacrifice), because it levelled the difference between the true victims and those perpetrators who willingly decided to sacrifice themselves for the Third Reich—an argument that he reiterated during the Holocaust Memorial debate (Olsen 2014: 284, 295–6). For such a political reading of the monument, see Olick (1999) and Nive (2002: 195).

[15] It must be emphasized that there was nothing unprecedented in the presence of enemy soldiers on Italian soil. After the First World War, Austrian–Hungarian soldiers were laid to rest in Italy and a large part of the German dead were buried in foreign countries (e.g., Belgium, France, Poland, the Soviet Union, and especially France). The main actor involved in the reburial policy concerning the German war dead was the Volksbund Deutshe Kriegsgräberfürsorge established in 1919.

images of soldiers as victims of war. As a result, individuals have upstaged the nation they belong to and, more importantly, the cause for which the country fights. In this way, the individual seems to have achieved a final triumph over the nation and the war cause for which soldiers are expected to be sacrificed. While in the previous approach to commemoration there is an explicit link between the fallen and the cause for which they sacrificed their lives, such a link is now either absent or secondary. Further examples of this attitude to war commemoration are too numerous to present in much detail, but, as we shall see, nowhere is this approach crystallized more effectively than in the Vietnam Veterans Memorial, whose importance is worth discussing at length.

The Vietnam Veterans Memorial: The Triumph of the Individual over the Cause

The Vietnam Veterans Memorial in Washington is spatially located between the Lincoln memorial and the Washington monument, and politically situated at the centre of American history. The project was sponsored by a veterans' association stemming from the initiative of Jan C. Scruggs, a 29-year-old Vietnam veteran himself who had the idea after watching Michael Cimino's *The Deer Hunter*. In its original version, the Vietnam Veterans Memorial consisted only of the Memorial Wall, which is formed of two black granite walls set into the ground, on which the names of the American fallen are inscribed, both the killed and the missing in action.[16] The names are listed in chronological order by date of casualty; no details of rank, unit, or function are provided.

Maya Lin, the architectural designer of the Memorial, declared that her project was inspired by First World War memorials, especially by the Memorial to the Missing of the Somme at Thiepval, which bears the names of thousands of British soldiers with no known graves. However, there is a striking difference between First World War monuments and the original version of the Vietnam Veterans Memorial.[17] In the latter, there was no celebration of patriotism, no celebration of heroism under arms, no celebration of traditional warrior values. There were only names and no reference to the war. The

[16] The wall includes more than 58,000 names, though it originally listed 57,939 names. There is now also a virtual Wall of Faces, which allows family and friends to share memories and post pictures of the American fallen of the Vietnam War. See https://www.vvmf.org/Wall-of-Faces/.

[17] Actually, the Vietnam Veterans Memorial recalls the monument to the partisan women of Villa Spada in Bologna, Italy, inaugurated in 1975 and formed by a wall with a row of red bricks bearing the names of 128 women who died for freedom. For an overview of this memorial—though making no explicit parallel with the Vietnam Veterans Memorial—see Dogliani (1995).

Vietnam memorial, at least in its original version, was a purely funerary monument that, by showing only the names and dates of death of the fallen, called attention only to grief without presenting any visible justification for the war. Both the choice of using black granite instead of white marble and the horizontal rather than vertical alignment—what Marita Sturken (1991: 123) described as the 'antiphallic form of the memorial'—placed the emphasis on human loss rather than on glory and patriotism. The old triumphalism of First World War memorials is wholly absent from this version. Neither do we find any signs of the traditional clichés of masculinity, power, and nationalism. In a way this memorial might not even be regarded as a war memorial at all. As Nathan Glazer (1996: 27) noted, the memorial 'does not tell us that these men died for their country, or for liberty, or for democracy, or even that they died in vain. It says nothing except that they died.'

Not surprisingly, the purely funerary nature of the Vietnam Veterans Memorial was interpreted as questioning war itself and created a torrent of criticism: 'a tribute to Jane Fonda', 'an open urinal', a 'perverse prank', 'something for New York intellectuals', and the like (Wagner-Pacifici and Schwartz 1991: 394). In more sophisticated vein, the editors of the conservative journal *National Review* (1981) argued that the 'mode of listing the names makes them individual deaths, not deaths in a cause'. In their ideological battle against the monument and unlike the vulgar detractors of the memorial, these conservative critics grasped the truly revolutionary shift in war commemoration marked by the Vietnam Veterans Memorial: the passage from death in war as a glorious, collective sacrifice for a national cause to an understanding of death in war as a politically meaningless and private tragedy.

Political pressure forced the US Commission of Fine Arts to modify the original project. In 1984, two further components were added to the original structure: an American flag and a statue of three soldiers (*Servicemen*) returning from patrol. With these additions, not only death was represented, but also traditional masculine images of the nation and combat (Hass 1998: 18). These features attempted to shift attention from the dead to conventional commemorative themes such as nation and sacrifice. After the alterations, President Reagan—who had originally refused to take part in the inauguration of the monument and had sent an unknown official to the ceremony—officiated at the unveiling of the flag and statue on 11 November 1984.

Such a return to First World War-like commemoration does mark a certain departure from what appeared to many as the pacifist bias of the Vietnam memorial. However, even the second, final version does not imply a true revival of the patriotic certainties of the past and conveys no overtly didactic message. First, the soldiers are not in the attacking posture of many First World War memorials. They are simply advancing over the open field looking at the wall (that is, at the dead); there is neither a depiction of active combat nor a representation of potential violence. Secondly, the three soldiers have been placed at a distance from the wall of names, and not at its apex, as Frederick Hart, the sculptor of the *Three Servicemen*, suggested and as was fairly common in First World War monuments. Soldiers and weapons might still be portrayed but were no longer at the centre of a memorial. Finally, though the Vietnam Veterans Memorial underwent several changes, only one feature stayed the same: the complete inventory of the (American) war dead.[18] The names of the fallen soldiers are still the central subject of what remains largely a funerary monument. Indeed, the Vietnam Veterans Memorial still looks more like a *memento* rather than a monument and, even in its final version, it fails to answer Norman Mailer's well-known question (1967): 'Why Are We in Vietnam?'

Even the way visitors approach the memorial reveals the prominence of the dead over the cause. It is noteworthy that, on inauguration day November 11, 1982, the most solemn aspect of the ceremony had already taken place at the National Cathedral on the 10th, when the names of the war dead had been read out in a 56-hour candlelight vigil (Wagner-Pacifici and Schwartz 1991, 378). Furthermore, anyone who has been to the memorial knows that most visitors come to see and touch the names of the fallen, and not to see the American flag or the patrolling soldiers. The wall of names remains the most important architectural element and the one with the most emotional significance for visitors.[19] Finally, it should be noted that the Vietnam Women's Memorial, dedicated on 11 November 1993, reinforces the tragic meaning attaching to death in modern war memorials. Formed of three women, one of whom is tending a wounded male combatant, the monument depicts only the violent result of war and not any potential positive meaning.[20]

[18] Non-western deaths are, indeed, rarely mourned or commemorated by western societies. The absence of commemoration for the Vietnamese, Laotian, and Cambodian dead in the Vietnam Veterans Memorial is quite telling in this respect (Hass 1998: 151), especially since the war dead of this conflict are estimated at more than two million.

[19] According to the National Park Service, the Vietnam Veterans Memorial is the most visited war memorial in the United States.

[20] Current attitudes towards soldier deaths, and most likely attitudes towards death in general, are nicely captured by Simone de Beauvoir (1966: 106) in a passage of her moving book *A Very Easy Death*, a work dedicated to the final weeks of her mother's life: 'There is no such thing as a natural death: nothing that ever happens to a man is ever natural, since his presence calls the world into question. All men must die but for every man his death is an accident, and, even if he knows it and consents to it, an unjustifiable violation.'

Interpreting the Memorial: Soldiers Are Mourned rather than Celebrated

The revolutionary shape of the Vietnam Veterans Memorial has generated debate among students of memory. According to some scholars, its shape is the outcome of a difficult task: how to commemorate a burning defeat. On this point, James Mayo (1988: 70) argued that memorialization of military defeat is different from commemoration of victory: 'defeat ... cannot be forgotten and a nation's people must find ways to redeem those who died for their country to make defeat honourable. This can be done by honouring the individuals who fought rather than the country's loss cause.' Employing a different argument, in an oft-quoted article, Wagner-Pacifici and Schwartz (1991) suggested that the Vietnam Veterans Memorial expresses the contradictions of American society in respect of the Vietnam War and, in particular, the problems involved in commemorating a divisive and unpopular conflict: 'The Memorial as it is presently constituted by wall, flag, and statute remains a multifocal version of the war monument genre, a version ambiguous enough to accommodate a wide span of commemorative meanings' (Wagner-Pacifici and Schwartz 1991: 408). Accordingly, the memorial embodies the extreme discord that divided American society with respect to the war.

These two interpretations concur in putting the final shape of the memorial down to the trauma of Vietnam. Although there is more than a grain of truth about such readings, they all treat the Vietnam Veterans Memorial in isolation, preventing any understanding of its meaning in transnational and historical terms. On the other hand, if the Vietnam memorial is examined from a long-term perspective, it can be read as the latest stage in the historical development of war commemoration. In order to understand what was really new about the Vietnam Veterans Memorial, as opposed to what *seems* to be new, the historical perspective offered in this research appears useful.

No doubt, the Vietnam War differed from previous conflicts because it was perceived by a large section of American society as an unnecessary defeat whose cause and final result could not be commemorated in ordinary ways: hence the centrality of the war dead. Yet, Chapter 3 has shown that a defeat and even a divisive military catastrophe could be celebrated by referring to its cause. After the Franco-Prussian conflict (1870), the French responded to their humiliating rout by creating a patriotic discourse centred on hundreds of memorials that stressed heroic defence and resistance. Moreover, after the First World War, commemoration focused on the fallen even in victorious countries. Last but not least, though Vietnam was a socially polarizing factor within American society, it should be noted that both France and Germany were in a state of civil war at the end of the Franco-Prussian conflict and First World War, respectively. Nevertheless, the rites of commemoration and remembrance did not call patriotism into question.[21]

By referring only to the dead without a noble cause for the supreme sacrifice, the Vietnam Veterans Memorial appears to mark the latest stage of development in war commemoration, which in the early modern period excluded the rank and file and now virtually refers only to them. As compared with memorials dedicated to previous wars, a characteristically different set of beliefs appears to be at work in this memorial. While the original cult of the fallen was closely linked to the war cause, it progressively lost its traditional meaning after the Second World War and is now primarily related to individual loss. Although the language of martyrdom for a noble purpose has not gone completely, it resonates less today than it did in the past (Underhill-Cady 2001: 172). The emphasis is no longer on the military qualities of combatants but rather on their human character. The main thing that remains is a name (or a face, especially on the dematerialized memorials on the Internet, see Grider 2007), and the main thing the name represents is a life that came to a premature end.[22] Thus, the Vietnam Veterans Memorial and other recent memorials should be understood not as a peculiar way to commemorate war, but rather as the latest chapter in the ample narration of attitudes towards soldier deaths in the modern period.[23]

The Korean War Veterans Memorial and the World War II Memorial

Unlike the Vietnam Veterans Memorial, in the Korean War Veterans Memorial, dedicated in July 1995, the cause is celebrated by a wall with photographic material and information about the conflict, two inscriptions explaining why the war was fought, and by nineteen statues of armed soldiers moving forward in V-formation and ready for action. Likewise, in the World War II

[21] It should also be noted that it was the Civil War (not Vietnam) that witnessed the harshest public opposition and violence against the draft in American history (Piehler 1995: 166). In the New York City drafts riots, 13–16 July 1863, 120 people were killed and an estimate of 2,000 were injured.

[22] Indeed, the war dead now have not only a name, but often a face. The increasing visibility of the war dead is paradoxically 'intensified by the fact that casualties are low enough to be reported about individually' (Ben-Ari 2005: 654).

[23] For a different interpretation of the Vietnam Veterans Memorial, which reads the emphasis placed on the dead as a symbolic strategy employed to deflect 'the troubling aspects' of the Vietnam War, see Hagopian (2009: 405).

Memorial, dedicated in May 2004, the conflict is commemorated through conventional patriotic devices (for example, statues of soldiers, military service seals, panels depicting the mobilization of American society, granite arches representing victory, and so on), which appear to subordinate the individual soldier to the nation. Although such an approach seems to contradict the post-Second World War 'negative' culture of commemoration that culminated in the Vietnam Veterans Memorial, these two memorials are not monuments to the beauty of dying on the battlefield. The scenes of warfare and potential engagement represented in the memorials provide a descriptive account of the war effort, but they do not portray war as a glorious killing enterprise. The classical architecture of the World War II Memorial may appear at first glance as a tribute to the American nation (Doss 2008), but there is none of the exaltation of martyrdom typical of First World War monuments. Moreover, the traditional shape of the Korean War Veterans Memorial is explained by a simple fact: unlike previous war memorials in the United States and in Europe, societal groups were totally excluded from deciding for or against its final design. The project of the Korean memorial was kept under strict governmental supervision; control over the design was conferred on an Advisory Board formed of twelve Korean war veterans appointed by President Reagan. Moreover, congressional legislation and monument competition instructions were precise in specifying what the memorial should and, above all, should not look like. Not only did the Advisory Board refer negatively to the Vietnam memorial several times, but, in the instructions for the design competition, it explicitly defined as unacceptable those projects that had an inherent 'essence of grief' (in Schwartz and Bayma 1999: 952). While the Vietnam Veterans Memorial was the outcome of negotiation between society and the government (the former played a critical role in raising financial resources and influencing its design and construction), the Korean War Veterans Memorial was from beginning to end an entirely state-controlled project that excluded civil society from the decision about how the war should be visually commemorated and remembered. In other words, it is a totally politicized memorial where popular attitudes towards the war dead have been silenced. Although the government certainly succeeded in deciding its final shape, the analytical value of the Korean memorial as meaningful evidence of current attitudes towards soldier deaths thus needs to be discounted. Having been under strict government control, it can actually be considered as an attempt *to lock the stable door after the horse has bolted.*

Corpses: Leave Nobody Behind

The treatment of the earthly remains of the war dead in the post-Second World War period provides further evidence for the shift in attitudes towards soldiers' life and death. As a matter of fact, not only has the meaning of death in war changed, but also soldiers' corpses have acquired a different value, becoming the object of a true obsession. On the one hand, there is a sort of agreement that, if someone is missing, comrades will attempt to rescue them. On the other, great efforts and costs are devoted to retrieving and identifying the remains of the fallen soldiers.

As considered in Chapter 3, recovering the bodies of the war dead—a policy pursued by the Union during the Civil War—has also been an important issue for the Prussians since the 1860s, and for virtually every European country since the First World War. Burial policies during and in the aftermath of the Second World War were also informed by a similar concern. In this conflict, to a considerable degree, western countries employed the rituals and organizations created for the First World War in the disposal of their fallen. Obviously, there were improvements in the management of disposal policies, and many new cemeteries were created in the various theatres of war, but no great alteration of approach took place. Even the United States, whose casualties were 5.4 times higher than they had been during the First World War (Congressional Research Services 2019, 2), devised no innovative ways of dealing with the war dead. As the concentration of corpses proceeded, the temporary cemeteries, which had been built to start with, were either closed or transformed into permanent burial sites. Several large graveyards were established overseas for the concentration of the dead in 'friendly soil' such as France, Italy, England, and the Philippines. Finally, families were given the last word as to repatriation of their loved ones, though the government had decreed that graves should be marked by uniform tombstones (Piehler 1995: 130–1).

A similar policy was pursued in France, where a 1946 law required armed forces to offer free delivery and transportation of corpses to the families. But here we can record a significant change in the relation between the living and the dead. In the aftermath of the Second World War, as Philippe Ariès (1981: 556) noted, 'the French people refused to turn their dead soldiers over to the large national cemeteries like those of World War I; they preferred to keep them in family graves'. Not only did such a decision show an increasing attachment by families to their loved ones, but it also revealed a relative weakening of the collective narrative and glorification surrounding the fallen, as already described. French families made it explicit that the dead belonged to them rather than to the state. A similar development took place in Italy, where, after the war, the vast majority of the families of the fallen requested the return of their loved ones' bodies (Schwarz 2003: 560–1).

The treatment of the human remains of the war dead in the post-Second World War period attests to this significant shift. As compared with burial polices of previous conflicts, a characteristically different set of beliefs appears to be at work in the decades following the Second World War. In particular, while the original cult of the fallen was closely linked to the idea of glorious death, it now relates primarily to the human loss of the families. The emphasis is no longer mainly on the collective sacrifice for the country, but rather on the individual life that came to a premature end.

The Increasing Importance of Corpses

While the main concern during the American Civil War and the two world wars had been with marking the plots of fallen comrades in temporary cemeteries, since the wars in Korea and, especially, Vietnam a true obsession with leaving no body (dead or alive) behind has emerged. In Korea, the United States abandoned battlefield burials after December 1950 and adopted the policy of Concurrent Return: the war dead were not to be buried in provisional grounds, but concentrated at collection points and then sent to Kokura, Japan, to be identified and eventually returned home (Keene 2010: 66). In order to prevent mistaken identification, the process of recovery was conducted as rapidly as possible, which made it possible to name roughly 97 per cent of the recovered American war dead of the Korean War (Anders 1988),[24] though more than 7,500 Americans remain still unaccounted for.

A similar course of action was adopted in Vietnam, where, thanks to improvements in equipment and logistics, such as the use of helicopters, the bodies were recovered soon after death, embalmed in mortuaries located in Da Nang and Tan Son Nhut, and sent home within seven or ten days after death (Sledge 2005: 57, 80).[25] Only 1,584 of the American soldiers fallen in

[24] The war dead of the other UN warring states were temporarily buried in the United Nations Memorial Cemetery in Busan, South Korea. In the following years many countries, such as Belgium, Colombia, Ethiopia, Luxembourg, the Philippines, and Thailand, repatriated their fallen soldiers.

[25] Being able rapidly to reach the wounded within the battle area and transport them to military hospitals, helicopters, like the Bell UH-1, also proved to be true life-saving vehicles. It should be noted

the Vietnam War are still unaccounted for, a relatively small number, bearing witness to the great efforts made by the federal government to recover and identify its war dead. Since the Vietnam conflict, the US military have organized an even more sophisticated system for retrieving the bodies of the fallen and returning them to the next of kin (Samet 2005).[26] Moreover, thanks to the introduction of dental records, DNA analysis, and other sophisticated techniques, the process of identification has become more and more effective, almost an exact science.

In 2003, contrary to the traditional policy of burying the war dead in the country where they fell, Britain also decided that all service personnel who died during military operations abroad should be repatriated at the government's expense (Summers 2010: 51). British families are also allowed to choose whether to have the Ministry of Defence service headstone or to opt for a personal design, which testifies to the fact that not only the natural bodies of soldiers but their social bodies as well are now owned by the families rather than the government. Thus, by allowing families to personalize the burial of their loved ones, national worship becomes secondary to the private mourning of single individuals.[27]

Not only do western countries recover the dead of current wars, but, since the mid-1970s, great efforts have also been made to search, retrieve, and name the unaccounted-for from past wars. In the United States, in order to carry out this policy, technically termed 'historical recovery' (Sledge 2005: 82), the Central Identification Laboratory Hawaii (CILHI) was created and tasked with this new mission in 1976.[28] In Britain as well, identifying the unnamed of past wars has become an important, sensitive issue both for the government and for private individuals. In 2009, during the largest operation undertaken in the post-Second World War period by the Commonwealth War Graves Commission, the remains of 250 British and Australian soldiers

that, thanks to the use of aircraft, evacuation capacities were also improved in the course of the Second World War.

[26] For information and numbers on the Americans who are unaccounted for in South-East Asia, see https://www.dpaa.mil/Portals/85/Statistics%20as%20of%20March%201.pdf. Activities of the so-called casualty branch, an office of experts in handling military death, include not only the recovery and identification of the war dead, but also the organization of funeral and memorial services as well as personalized assistance to families to meet particular requests.

[27] Such an individualization process can also be seen at work in Israel. Since the beginning of the 1990s, tombstones have been allowed to express personal information and the individual identity of the fallen. Such a change resulted largely from pressure by the families of the fallen against the previous view that stressed the importance of uniformity as a mark of the unity of purpose between the fallen and the Israeli nation (Katz 2014 401–9).

[28] In 2003, CILHI and the Joint Task Force-Full Accounting (JTF-FA) merged into the Joint POW/MIA Accounting Command (JPAC).

were exhumed near the village of Fromelles, France, from an unmarked mass grave dug by the Germans in July 1916. The human remains were later reinterred close to the original burial site in a newly created military cemetery. Thanks to DNA tests, a number of the dead were positively identified, including seventy-five Australian soldiers (McPhedran 2013).[29] In Italy too, historical recoveries are taking place. For example, in June 2018, a hundred corpses (ninety-four unnamed and six positively identified) of soldiers fallen in Russia during the Second World War were returned, thanks to operations organized by Onorcaduti, a special office of the Italian Ministry of Defence.

Devotion to war dead will appear to be even more remarkable if one considers that countries invest significant administrative, diplomatic, and financial resources in historical recoveries. The US government, for example, spends approximately $100 million annually on the search for its missing (Allen 2009: 2; Keene 2010: 75). Although recovering, repatriating, and identifying the remains of the war dead is a costly scheme, it is regarded as a sacred obligation, and, as such, it justifies all efforts.[30] Furthermore, despite the common use of the word 'bodies', what is often at stake here is just tiny fragments of corpses, such as portions of skeleton and teeth (Hawley 2002: 50), which bear little resemblance to a person but are nevertheless grieved over by the families (Wong 2005). In this respect, the story of the American Unknown Soldier of the Vietnam War is emblematic of the relationship between the living and the war dead in the contemporary western world.

Entombed as unidentified in 1984, the remains of the soldier 'known but to God' were disinterred in May 1998 after 'forceful pleas' by an American family who believed that the Unknown was their loved one shot down in Vietnam in 1972 (Wagner 2013: 641). The remains were then identified as belonging to First Lieutenant Michael J. Blassie and, finally, transported to his surviving relatives in Saint Louis, Missouri. Although, as Wasinski (2008: 119) and Wagner (2013: 647) rightly maintain, the traditional narrative of death in war as the ultimate sacrifice for the country is far from being gone—Blassie's remains were reinterred with full military honours at Jefferson Barracks National Cemetery[31]—in this family's successful struggle to take back their loved one

[29] Even private individuals are involved in historical recovery. In 1988, Brian Oldham began to work on the remaining records of the corpses buried in the Haidar Pasha cemetery with a view to compiling the great Crimean War index (Summers 2010: 11).

[30] Such a position, it is worth noting, finds bipartisan support from both Democratic and Republican administrations (Wagner 2015:167).

[31] The presence of a variety of attitudes and values attaching to the corpse of the fallen is not surprising. As Verdery (1999: 28) explained, human remains are 'concrete, yet protean; they do not have

from the state we note further evidence of the shift away from a sacrificial notion of death in war to an emphasis on the pain and suffering borne by the deceased and their families.

Doubtless advances in the technology of identification have made it possible to enact this type of individuated remembrance. But technology does not fully account for the unprecedented care towards the bodies of the fallen. The example of the Vietnam Unknown's identification and the practices just discussed attest also to a deeper change in the societal meaning attaching to soldier deaths: from the collective cult of national heroes we have moved towards a more individualized understanding of death as a private tragedy.

This is also apparent if one looks at military obituaries on the Web, which often report not only information on how death occurred, but also biographical details of marital status, family situation, hobbies, and past hopes of the dead (King 2010). These obituaries construe the soldiers as irreplaceable human beings and are meant to convey the uniqueness of the dead. In the telling words of a British widow commenting on her decision to issue a biographical obituary on her husband killed during Operation Telic (Iraq): 'I have decided to issue this statement because I feel strongly that I should make clear that Matty wasn't just another number added to a casualty list' (in Zehfuss 2009: 430). The emphasis is no longer on the glorious sacrifice for one's country, but rather on the personal traits of the war dead.[32] Although soldiers are still commemorated as a special category, different from civilians, it is no longer and not only a tale of states and armies, but rather one of individual people with stories to be told.[33]

a single meaning but are open to many different readings'. However, what we are registering is a shift in the balance of meanings. While the state's attempt to appropriate the death of soldiers persists, at the societal level their death is primarily interpreted not as a beautiful act of sacrifice for the nation, but rather as a tragic human loss. Moreover, many US soldiers who died abroad are interred in their hometown cemeteries (Keene 2010: 60).

[32] For a different argument stressing the continuing importance of 'national kinship' as the main way to justify military deaths, see Åse and Wendt's analysis (2017) of media coverage of Danish and Swedish military losses in Afghanistan.

[33] In his analysis of the obituaries of the British dead at Helmand, Anthony King (2010: 10) concludes that the war dead are not mainly understood to have died for their country or for 'wider national' purposes, but are rather considered as 'personalities, defined through their unique professionalism'. Although similar, King's argument differs from the one expressed in these pages in one significant aspect. According to the British sociologist, military deaths are understood 'as the meaningful expression of a man who defined himself by his profession'. Death is tragic, as I argue as well, but far from being pointless because it is the outcome of a chosen professional vocation that embraces mortal danger. I have no doubt that the war dead are still remembered as soldiers rather than civilians, but emphasis on professionalism is found mostly in military obituaries published on the Ministry of Defence websites rather than in local and national papers.

No Longer Expendables: Casualty-Averse Warfare

During the First World War, a military padre in France maintained that the 'soldier's business is to kill the enemy ... and he only tries to avoid being killed for the sake of being efficient' (in Bourke 1999: pp. xiii–xiv). This is certainly an accurate description of how commanders employed soldiers for much of the modern epoch. However, such a martial characterization of combatants does not describe western warfare in the current world. The new perception of soldier deaths in war that has just been highlighted with reference to commemorative and disposal practices seems to have affected how western countries employ their soldiers on the battlefield as well. While death in battle used to be either glorified or devoid of any social value, present-day governments and military authorities are determined to avoid it, thus turning the cult of the fallen soldier into an attempt to foster the preservation of combatants' lives. In so doing, the sanctity of bodies (at least of western bodies) has become a touchstone of the western way of war.

What is striking in current western warfare is the extent to which armed forces try to minimize friendly casualties. Soldiers are no longer considered as mere instruments for achieving a particular military and political end. In the process of killing enemies, western states try their best to prevent their combatants from getting killed, and, when soldiers are wounded, they are immediately recovered and medically treated. The death of soldiers—rich and poor—is seen now as an intolerable social loss. Reversing Napoleon's dictum, one may easily note that soldiers *are no longer made to be killed*.[34] The point is that individual life is deemed to be valuable beyond a soldier's fighting potential, whereas in the past soldiers were employed with caution only for the sake of preserving the army.

Aversion to friendly casualties has turned into policies that have, at the micro-level of military practice, established the protection of troops as a major mission in combat operations as well. While the number of wars waged by western states shows that casualty aversion does not seem to have reduced the proclivity towards the use of force abroad, it has, nonetheless, had two main effects. On the one hand, casualty sensitivity appears to jeopardize states' willingness to continue fighting when fatalities mount, especially in peacekeeping operations and humanitarian interventions. For example, the Beirut barracks bombing on October 1983, which killed 299 American and French military personnel, led to the withdrawal of the peacekeeping Multinational

[34] According to Michael Walzer (2006: 136), Napoleon once said: 'Soldiers are made to be killed.'

Force from Lebanon in the following months (Moskos 2001). Ten years later, the American decision to withdraw its armed forces from Mogadishu, Somalia, in October 1993 after the killing of eighteen service members supporting the peace operation UNOSOM II was interpreted as the first case of casualty aversion in the post-cold-war period. The footage showing the dead bodies of US Rangers being dragged through the dusty streets of the Somali capital seems to have had a powerful impact on the decision to abandon the country. Casualty sensitivity also appears to have led to the decision to withdraw UN forces from Rwanda in 1994, after the killing of eleven Belgian soldiers, a decision that turned into a death sentence for hundreds of thousands of people. Furthermore, in May 1994, when massacres were taking place in Rwanda, the Clinton Administration issued a number of restrictive guidelines on humanitarian intervention—that is, Presidential Decision Directive 25 (PDD-25). Even though the directive was drafted several weeks before the Mogadishu disaster (Gasbarra 2017), the events in Somalia reminded Washington of the potential human cost of peacekeeping.

On the other hand, casualty sensitivity has shaped the ways force and troops are employed on the battlefield. In particular, aversion to western casualties has developed into three main military policies that have made the protection of troops a major mission in combat operations too: a preference for waging war from a *safe distance*, reliance on indigenous forces during land operations, and, finally, the adoption of force protection measures to safeguard soldiers' lives when the use of western ground troops cannot be avoided.

The reliance on air power, typical of current interventions, can actually be traced back to the early 1950s. Since the Korean War, western states have relied heavily on air power in order to reduce military fatalities. While the war in Korea is often described as an infantry war, it should be remembered that, when stalemate was reached in 1951, ground forces adopted a strategic defence position, and air power became the main offensive weapon. As Adrian Lewis (2007: 84) argued, the 'citizen–soldier Army of the United States would never again fight a major war with offensive strategy and doctrine. In 1951, major limited wars came to mean a strategically defensive ground war in which the Army was not supposed to produce victory.'

The attempt to spare American lives was a major priority during military operations in the Korean peninsula. On this point, General Van Fleet, Commander of the Eighth Army, declared: 'We must expend steel and fire, not men. I want so many artillery holes that men can step from one to the other' (in Toner 1981: 84). Owing to concerns over mounting American casualties, President Eisenhower declared that the war should be turned into a conflict of

'Asians against Asians': 'our boys do not belong on the front lines' (in Pearlman 2008: 253). Historian Theodore R. Fehrenbach (1994: 151), who fought in Korea, draws a good outline of the leading strategy in America's conduct of the war: 'American units were proceeding to destroy utterly enemy-held towns and villages rather than engage in the costly business of reducing them block by block with men and bayonets ... If bombing and artillery would save lives ... saving lives obviously had preference.' As the war progressed, General Ridgway (1967: 222) was also more and more concerned about expending American combatants. The final result of this military approach was described by the Korean ambassador to the United States as the transformation of the Peninsula into a 'gigantic charnel-house' (in Toner 1981: 85).

Not surprisingly, since the Korean War, air power has gradually become the favourite instrument of US offence strategy, as American warfare in Vietnam clearly testified. Although nearly 60,000 American soldiers ended up dead in Vietnam—a great human toll, which shows how the US military and policymakers were still ready to sacrifice a high number of lives in that epoch[35]—one may describe US strategy in that conflict as characterized less by ground operations than by the use of air power and heavy artillery. As General Palmer explained: 'maneuver elements found the foe while firepower eliminated him ... When contact was made, American units, preoccupied with avoiding casualties, generally fell back into a defensive perimeter to call for air and artillery' (in Lewis 2007: 260). Not by chance, the US Air Force dropped seven million tons of explosives all over Vietnam, roughly five times the amount discharged by Anglo-American forces during the Second World War (Harrison 1993: 133). There is also evidence that General Westmoreland's stress on firepower was an attempt to replace men with machines and, in so doing, minimize friendly casualties (Krepinevich 1988: 164). Furthermore, because enemy and friendly casualties became a measure of success in the course of the war, military doctrine tended to prioritize tactics that ensured a positive body count. Interestingly, this novel western sensitivity to casualties emerged in spite of the fact that, as in the past, 'the bulk of ground combat forces ... was mainly drawn from lower socioeconomic groups' (Moskos 1970: 10).

The reliance on air power for reducing friendly casualties is even more evident during conflicts in the post-cold-war period.[36] It is no coincidence that in every major intervention recently involving western countries, from Bosnia

[35] It must be noted that the North Vietnamese were not casualty averse at all. As General Giap brutally put it: 'Every two minutes three hundred thousand people die on this planet. What are forty-five thousand for a battle? *In war death doesn't count*' (in Fallaci 1976: 76; emphasis added).

[36] The next few paragraphs are based partly on Castelli and Zambernardi (2017).

to Libya, the use of long-distance missiles launched by land, sea, and air has played a prominent role. In operations such as Deliberate Force (Bosnia and Herzegovina), Infinite Reach (Afghanistan and Sudan), Desert Fox (Iraq), Allied Force (Kosovo), and Unified Protector (Libya), the exclusive reliance on air power made it possible to keep western combatants actually outside the war zone.

While it is a fact that the United States provided most of the firepower employed in these bombing campaigns, European countries, too, seem to have become perfectly comfortable with this military approach. And, while NATO's European members at the time of the war in Kosovo—unlike the United States (Caniglia 2001; Feaver 2003: 274–6)[37]—were willing to launch a ground operation, they moved closer towards a preference for aerial bombing in the following years, as clearly shown by the way they fought in Libya in order to stop Gaddafi's operations against civilians.[38] Although a variety of motives may account for the use of purely aerial strategy in these interventions, one of the main reasons seems to lie in the simple fact that neither the United States nor European countries were ready to pay the human cost of ground warfare. As unambiguously stated in the *Kosovo/Operation Allied Force After-Action Report* to the United States Congress, the 'paramount lesson learned from Operation Allied Force is that the well-being of our people must remain our first priority' (US Department of Defense 2000: 108).

The possibility of putting the living flesh of soldiers out of harm's way is also one of the principal reasons for the increasing use of drones. Drones, it is true, are often used in close-support missions to back up local forces engaged in ground warfare, as in Libya during *Operation Odyssey Lightning* (Turse et al. 2018); but, like ballistic missiles and all types of remote warfare, drones avoid direct relation between those firing the weapon and the target. While there are certainly a number of reasons why drones are increasingly being employed—for example, they allow continuous surveillance of targets, they enable identification of potential suspects, and they are also cheaper than other weapons systems (Horowitz 2017; Boyle 2020)—one of the main reasons for their use lies in the fact that these remote-controlled vehicles are crewless,

[37] Based on an estimate provided by General Wesley Clark of 117,000 casualties (in Der Derian 2009: 190)—a number that was most likely inflated—the Clinton Administration utterly ruled out the plan for a ground operation against Serbia and opted for attacking from afar, which made it possible to have no casualties. Moreover, in order not to be targeted by Serbian anti-aircraft, American pilots were forbidden to fly below 15,000 feet.

[38] Recalling the traditional warrior ethos of the military, French general Philippe Morillon, former commander of the UN Protection Force in ex-Yugoslavia (1992–3), asked 'how can you have soldiers who are ready to kill, who are not ready to die?' (in Pfaff 2004: 132).

and thus no one risks his or her own life. As an Air Force officer put it: 'The real advantage of unmanned aerial systems is that they allow you to project power without projecting vulnerability' (in Chamayou 2015: 12).

Removing troops from the battlefield by means of air power is not the only way to limit casualties. The effects of casualty aversion on military practice not only include the use of long-range bombing, but also affect the style of land warfare. For the current western way of war entails another typical aspect, which is of considerable interest. When land combat is inevitable, western states rely heavily on indigenous forces to carry out their military operations. Although the western states supply most of the heavy weaponry (bombers, artillery pieces, and so on), it is up to local actors to provide forces on the ground.

The policy of using local forces has been adopted in a variety of ways, depending on the geographical context: it was used in Vietnam (Vietnamization from 1969 to 1975) and, more recently, in Iraq and Afghanistan. Again, indigenous forces allied with western states played a major role in many post-cold-war conflicts: the Croatian army and Bosnian militias in former Yugoslavia, the Kosovo Liberation Army (KLA) against Milosevic's Serbia, Kurdish and Sunni militias in Iraq, Libyan rebels against Gaddafi's regime, and Peshmerga and Iraqi forces in the conflict against ISIS.[39]

Indigenous forces are advantageous, not only on account of their familiarity with the local culture and terrain, as often argued (Cassidy 2006; Galula 2006), but because they know how to put up with suffering that westerners are not ready to bear. From this viewpoint, the military operation in Libya (not its outcome) seems to be an 'ideal' pattern for western states and probably a model for future interventions as well: bombing from a safe distance and subcontracting the fight on the ground to local coalition partners and militias. Not only did indigenous militias play a critical role in interventions such as Kosovo and Libya,[40] but their military effort was also casualty free for western

[39] Use of indigenous troops in combat is not a practice introduced only in recent years, of course. European colonial powers always had their colonial forces—the British, Germans, and Italians had their *askari* soldiers, the French had their *tirailleurs indigènes*, and the Spanish their *regulares*. However, today's policy of relying on local forces is significantly different from that of the past. First, although local forces are advised, equipped, and trained by Americans and Europeans, they retain their own organizations, agendas, and interests, which are often quite divergent from ours. Second, indiscipline, infiltration, and corruption among the indigenous armies affect their size and the quality of their performance. In other words, local forces might save western lives (when they do not shoot their western trainers!), but they are not always the ideal instrument to achieve the political goals that westerners are fighting for.

[40] Interestingly for the present research, NATO did not deploy ground forces after the military intervention in Libya.

powers.[41] Commenting on General Franks's decision to employ the Northern Alliance against the Taliban at the beginning of the conflict in Afghanistan, US general Michael DeLong was absolutely explicit in stressing this point: 'from our perspective, it was still less than it would have cost to put even one US battalion on the ground. In that sense, it was a bargain—and more important, it would help keep thousands of US soldiers out of harm's way' (in Buley 2008: 115).

Finally, casualty minimization is also attempted when western troops are used on the ground. Deploying ground troops is not the same as employing them, exposing them to lethal danger. Soldiers can be both deployed but not employed, and employed in a variety of ways that minimize the probability of being targeted. As combatants are no longer regarded as cannon fodder, protection of them is not automatically suspended when they are sent into the war theatre.[42] The third main feature of current western warfare, in fact, is the use of force-protection tactics when waging war without being anywhere near the enemy is not an option. While this policy used mainly to apply to the defence of military installations and bases,[43] the wars in Iraq and Afghanistan turned force protection into a major approach in combat operations too.

Force protection is a blanket term covering all those measures that are taken to minimize ground forces' exposure to enemy fire and to keep casualties low. Above all, it implies a separation or at least a distancing between western ground troops—who live in relatively safe and well-protected compounds and military bases—and the civilian population, who are constantly at risk of being targeted by enemy combatants and insurgents. By limiting contacts with the

[41] During the war in Kosovo, for example, the KLA played a critical role, not so much in its direct confrontation with the Serbian army, but rather because KLA attacks compelled Serbian forces to fight in the open and, hence, succeeded in exposing Milosevic's troops to NATO air strikes. Thus, in the first sixty days' bombing, air strikes were ineffective 'against ground targets … But once the KLA began its attacks and forced the Serbian military out in the open … the success of air strikes rose exponentially' (Townsend 2000: 10).

[42] Despite Maya Zehfuss's argument (2009) about the paradox of exposing to mortal danger soldiers who are 'grievable', it must be noted that protection of western combatants is not automatically withdrawn when they are sent into the theatre of war.

[43] Following the Khobar Tower terrorist bombing in June 1996, where nineteen American service members were killed and hundreds injured, Secretary of Defense William James Perry (1996) released a report recommending the adoption of a radically 'new mindset and dramatic changes' in the way the United States protects its forces deployed overseas. By declaring the necessity to defend American global interests and responsibilities, Perry (1996) suggested putting force protection 'on an equal footing with other mission goals' in the deployment of American forces. As a result, the Department of Defense launched a Force Protection program in February 1997. Only a year after the project had been created, the director of the US Air Force security forces, Brig. Geneneral Richard Coleman, announced that 'force protection has become the Air Force's highest priority … conducting that mission is now as important as projecting our combat power' (in Gentry 1998).

local population, the possibility of surprise attacks, ambushes, and hostile 'hit and run' assaults is greatly reduced for western soldiers.

Force protection might also extend to major tactical devices regarding rules of engagement (for example, checkpoint protocols, targeting, urban combat) and decisions on weaponeering. In particular, it ensures rules of engagement regarding checkpoint protocols and urban warfare, which can lower the threshold for the use of force (T. W. Smith 2008). Moreover, while the American definition of weaponeering does not overtly include force protection, the reality of military practice shows that casualty sensitivity can lead to the use of weapons that are utterly incapable of distinguishing between civilians and enemy combatants (Kaempf 2018). For example, American troops during the siege of Falluja in November 2004 and the Israeli Defense Forces (IDF) during the Cast Lead operation in Gaza in December 2008–January 2009 justifed using white phosphorus munitions as a means to light up the battlefield and, more importantly, as a force protection measure to obscure their soldiers from enemy fire.

Significantly, force protection has also been included in International Humanitarian Law (IHL)— namely, in the current rules governing the conduct of military operations. In particular, IHL allows commanders to take into account friendly casualties when considering the appropriate means and methods for a particular situation. It allows the so-called principle of distinction between combatants and non-combatants—that is, one of the cornerstones of IHL—to be superseded. To put it bluntly, killing civilians is seen as an acceptable outcome if the objective is to protect the lives of military personnel. For instance, France and Britain presented reservations to Protocol I of the Geneva Conventions declaring that the presumption of civilian status in cases of doubt (Art. 52(3)) could not override 'a commander's duty to protect the safety of troops under his command'. When Australia ratified Protocol I, the government specified that 'the term "military advantage" involves a variety of considerations including the security of attacking forces'. New Zealand added a similar specification (in Fredman and Steinberg 2010: 14).[44]

The imperative of protecting troops is so central today that it also affects the numbers of troops used in a specific military operation. NATO, for example, plans operations according to the possibility of providing the wounded with 'surgical care within the "golden hour" of sustaining wounds'

[44] The United States signed the two Additional Protocols but without ratifying them. Israel never signed the Protocols.

(Fazal 2014: 110).[45] In other words, the decision as to the number of troops to be employed not only reflects political and military exigencies, but also results from the goal of saving the potential wounded.

Finally, it should be noted that behind many ongoing and future developments in the military industry there lies a force protection logic. Automation of warfare through unmanned aerial vehicles (UAVs), unmanned combat vehicles (UCVs), and robotic weapons systems is the latest chapter in the attempt to remove soldiers from the battlefield (Singer 2009: 221, 226; Springer 2018). As one American official wrote in 2005, the whole rationale 'is that the President will wake up some day and decide he doesn't like the cut of someone's jib and send thither infinite numbers of Myrmidons—robotic warriors—and that we could wage a war in which we wouldn't put at risk our precious skin' (in Coker 2007: 119).

No doubt, the creation of human-like Myrmidons, through high-tech body armour like the now-defunct TALOS (Tactical Assault Light Operator Suit), might potentially solve the problem raised by casualty sensitivity.[46] However, though the speed of technological change is difficult to predict, post-human warfare appears many decades off, which means that in coming years western warfare is going to look like the military confrontations described in the final part of this chapter.

COIN in Iraq and Afghanistan

But what about the population-centric strategy employed by the United States and some of its allies in recent military interventions in Iraq and Afghanistan? Such a doctrine, known as COIN, which apparently relies on boots on the ground rather than firepower, seems at first glance to run counter to the trend described in this chapter (namely, the increasing removal of soldiers from the battleground). However, as I try to show in this section, the story is more complex than that.

In the initial years of the wars in Iraq and Afghanistan, the United States and other western states relied heavily on high technology, overwhelming air power, and small ground forces. Such a military approach was informally codified as the Rumsfeld doctrine, which promised military victory at moderate human and financial cost. In order to reduce risks, US and coalition troops

[45] The *golden hour* refers to the critical time (from a few minutes to a few hours) following a severe injury when medical treatment can prevent death.

[46] On the limits and fate of the TALOS program, see Cox (2014) and Douglas (2021), respectively.

rarely had contact with the local population, remained isolated in heavily fortified bases, and carried out rapid raids in armoured vehicles with frequent use of air strikes. However, after this strategy ended in failure to suppress the Iraqi insurgency, in December 2006 General David Petraeus and General James F. Amos supervised the compilation of a new counterinsurgency field manual—*US Army and Marine Corps Counterinsurgency Field Manual* (FM 3–24) (US Department of the Army 2006)—with a view to reconsidering military assumptions and doctrinal principles.[47] Hence, the difficulties of fighting an elusive insurgent in Iraq effectively roused the US military to the recognition that, against an asymmetric enemy fighting on its national ground, conventional methods were inadequate.

Precisely because victory in irregular warfare should not be equated with sustaining few casualties, the focus was moved from putting down the insurgency to providing security for the population. In so doing, the United States adopted what David Kilcullen (2006: 105) termed *the residential approach*: 'living in your sector, in close proximity to the population rather than raiding into the area from remote, secure bases.' To put it differently, with COIN the United States tried to transform its troops into *telluric* forces, who, by controlling the territory and having constant relations with locals, could establish a positive relationship with the indigenous population.

The doctrinal change, along with a series of political agreements with sections of the insurgency (that is, the Anbar Awakening), and the injection of additional troops provided by the so-called Surge in 2007, proved successful in reducing the level of violence in Iraq. Not only did the new approach result in lower violence and a reduction in the number of non-combatant fatalities, but, after an initial increase, the casualty rate for US troops declined as well (Goldberg 2014: 3).[48] Yet, despite such a positive development, Washington decided not to capitalize on the success and already by the end of 2007 had opted for

[47] To be fair to the American military, the so-called shift from counterterrorism to counterinsurgency can be dated to 2004, when, under the direction of General David W. Barno, population control in Afghanistan had become a primary goal, while General Mattis in Iraq restricted the use of artillery and changed the style of patrolling from armoured vehicles to foot patrolling. The first field guide for irregular warfare, stressing the principles of classic counterinsurgency, was released in 2004 (Buley 2008: 131). Moreover, not only was the 2006 *Manual* supervised by General Petraues and General Amos explicit in rejecting the initial strategy, but its previous 2004 version likewise emphasized the need to go beyond force protection: 'US forces have frequently resorted to fire power in the form of artillery or air any time they make contact. This creates two negatives in a counterinsurgency. First, massive fire power causes collateral damage, thereby frequently driving the locals into the arms of the insurgents. Second, it allows insurgents to break contact after having inflicted casualties on friendly forces' (US Department of the Army 2004: 3–10).

[48] One may argue that the success in stabilizing the country from 2007 to 2011 mainly resulted from political settlements agreed with Sunni insurgents, especially stemming from the Anbar Awakening—that is, the Sunni tribal uprising against al-Qaida in Iraq. Doubtless, it was not only the surge in troop

a gradual withdrawal until the final exit in December 2011. Military progress was not enough to justify continuation of the war effort. A combination of factors such as persistent pro-withdrawal sentiment within American public opinion, rising financial costs, and accumulating combat deaths (Mueller 2011) swayed President Obama in favour of the exit option. In other words, when a face-saving condition was reached with the temporary stabilization of Iraq, the United States decided it was time to withdraw from the conflict. Likewise, after the adoption of a similar COIN strategy in Afghanistan in 2010, Washington began to scale down the presence of its troops in the country, relying mostly on local forces and air power with an increasing use of drone strikes. Finally, President Biden (2021) decided to pull out troops, leaving Afghanistan in the hands of the Taliban and justifying the decision to end intervention by tellingly asking: 'how many more lives—American lives—is it worth? How many endless rows of headstones at Arlington National Cemetery?' Although Washington and its numerous allies still had the means to pursue such wars, they lacked the will to carry on with the fight. In both conflicts the weaker inflicted enough losses to make the stronger give up its military intervention.

Why did the United States and other western countries not carry on with the *telluric* strategy in Iraq and Afghanistan? From the perspective developed in this book, such a decision comes as no surprise.[49] The difficulty in giving up casualty-averse warfare, as this study has tried to show, is rooted in the priceless value that western societies attach to the lives of their soldiers. Despite the frequent claims by COIN proponents about rejecting measures that minimize risks to troops, the latter cannot be dispensed with because casualties are acceptable only if they are limited to low levels. Although soldiers were sent to places such as Iraq and Afghanistan to defend national interests, to fight international terrorism, and to protect human rights, once they had become engulfed in a difficult terrain, the minimization of human losses turned out to be the main objective. From this perspective, the great paradox of COIN lies in the simple fact that it is a doctrine that has been adopted in an epoch in which it cannot be fully implemented, because it implies risks and human losses that western societies are not ready to take anymore. The probable result is that future western and, especially, American warfare is going to look like the much-denigrated Rumsfeld doctrine, the so-called "light footprint"

numbers and the new strategy that contributed to stabilizing Iraq, but political agreements with portions of the insurgency. As Biddle et al. (2012, 10) rightly argued, 'a synergistic interaction between the surge and the Awakening is the best explanation for why violence declined in Iraq in 2007'.

[49] For a position that queries the adoption of a truly population-centric counterinsurgency, see Caverley (2016), who underlined that the ratio of manpower to firepower paradoxically decreased after the Surge in Iraq and in Afghanistan too.

approach based on precision firepower, special forces, and local partners rather than boots on the ground.[50]

Conclusion

Military policies meant to reduce friendly casualties have made it possible to avoid putting western military personnel in harm's way or at least to reduce casualties to extremely low levels. In present-day conflicts, the death tolls for western states have ranged from zero in Kosovo, where only air power was used and no combat fatality occurred, to a few thousands in Iraq and Afghanistan, which may sound like large numbers by today's standards, but is an extremely low human cost in historical terms.

Western leaders and, arguably, their citizens have found strategies and tactics that minimize casualties reassuring. Thinking that war can be waged at little human cost means believing that the world can be fixed through the use of force in a clean manner and with little bloodshed. Such a view is extremely convenient at a time when western states often find themselves employing coercive force abroad,[51]as it promises a way of coping with the costs of war to societies that increasingly tend to perceive soldiers as human beings whose lives should be preserved. It provides a simple and reassuring tactic for the conduct of war, with a promise of military success at tolerable human cost. While such policies represent understandable, noble efforts to reduce the human cost of war for western societies increasingly averse to incurring war dead, the final chapter of this research will illustrate why the current casualty aversion marks a further erosion of western military power.

[50] For a similar prediction, based on a different argument, see Gat (2011).

[51] This is far from being the legacy of the Bush Administration. In 2001, Andrew Bacevich and Eliot Cohen (2001: p. xi) correctly noticed that during 'his two terms in the White House, Bill Clinton, the first post-Cold War president, employed American military power more often, in more places, and for more varied purposes than any of his predecessors'.

5

Epilogue: The Further Erosion of Western Military Power

The man is the first weapon of battle ...

Charles Ardant Du Picq[1]

In a simple but enlightening article published in *Foreign Affairs*, meaningfully titled 'The Forgotten Dimensions of Strategy' (1979), Michael Howard reminded us that war is not a mere application of available technology but is in fact conducted in three further areas: the operational, the logistical, and the social. While no successful strategy can be formulated without taking into account all four dimensions, Howard suggested that one of these aspects might prevail over the others in some specific circumstances. Reflecting on the previous four decades of western warfare in developing countries, Howard (1979: 981) contended that it was the 'inadequacy of the sociopolitical analysis of the societies with which we were dealing that lay at the root of the failure of the Western powers to cope more effectively with the revolutionary and insurgency movements that characterized the postwar era, from China in the 1940s to Vietnam in the 1960s'.

Obviously enough, making policy without knowing and understanding one's enemy and its society is far from being an ideal strategy.[2] Yet at the root of the West's failure to cope with some of the present threats is not merely the incapacity to understand the peculiarities of other societies, but rather the failure to admit and overcome the sharp contradictions between the exigencies of war (namely, risking and losing lives) and our societal requirements (namely, minimizing casualties).

So far, students of international politics, philosophers, and jurists alike have been mainly interested in condemning the direct unethical consequences of the measures taken to reduce friendly casualties, in particular the fact that the use of air power and the adoption of force protection measures have shifted

[1] Quoted in Holmes (1989: 18).

[2] For a similar argument applied to the war in Afghanistan, see Farrell (2018). However, on the inherent limits and paradoxical outcomes of military interventions based on the attempt to understand alien societies, see Tripodi (2021).

Life, Death, and the Western Way of War. Lorenzo Zambernardi,, Oxford University Press.
 DOI: 10.1093/oso/9780192858245.003.0006

the burden of war from western soldiers to the civilians in the war zone (Kahn 2002; Shaw 2005; Der Derian 2009; Carpenter 2011; Kaempf 2018; Goozen 2021). However, a major political issue is also at stake here: since one of the most important means of war (that is, soldiers) cannot be used and sacrificed as freely as it used to be, war is becoming less and less useful as an instrument of foreign policy, especially in what Rupert Smith (2006) defined as conflicts 'amongst the people'—that is, in unconventional armed confrontations where the ultimate political objective is not to coerce individuals into submission but rather to capture the intentions and will of the people (that is, to create the conditions for a long-lasting political order). In this sense, the reluctance of both policymakers and military commanders to allow soldiers to take greater risks is not only an ethical failure but also a military and political problem.

It would be foolish to argue that western warfare has become ineffective. Actually, it is still very useful in destroying things, killing enemies, and overthrowing regimes. By assuring a high degree of accuracy, precision-guided munitions, drones, and other modern weaponry allow one to inflict immense damage upon the enemy while minimizing the possibility of being targeted in return. By no means coincidentally, western military technology was extraordinarily effective in toppling dictatorships such as Saddam's, Gaddafi's, and the Taliban's in a matter of weeks or months. From this viewpoint, the effectiveness of military power is not in dispute. However, it appears less adequate when it comes to converting military victory into a stable political outcome. In fact, being able to kill with near impunity and overthrow a government is far from ensuring a smooth transition to a new social order. If the ultimate end of war, as Clausewitz taught us, is to establish the conditions for a long-term peace,[3] it is inevitable to conclude that the results of western warfare have been profoundly disappointing in the post 9/11 era.

Whereas the great French historian Philippe Ariès (1981: 559) concluded his monumental study on western attitudes towards death by declaring that the desire for and the practice of denying death typical of modern society were 'robbing the dead and dying of all dignity', I would argue that, within the framework of armed conflict, the current social meaning ascribed to soldiers' lives and deaths is robbing warfare of part of its utility. Although pacifists may be content to see a decline in war's utility, this state of affairs is far from being a

[3] In his study on Clausewitz's philosophy of war—a book only partly translated into English—the great French intellectual Raymond Aron (1983: 97) clarified that, for Clausewitz, 'if strategy has one end, it could be summarized in a single word: peace. The end of strategy or of the conduct of the war is peace, not military victory.'

positive development, since this loss of effectiveness has not diminished the western proclivity for the use of force.[4]

An Increasingly Eroded Military Power

In his little but powerful book *The Blunted Sword* (1988), the late Evan Luard contended that military power had ceased to be the decisive factor in contemporary international affairs. He found it increasingly anachronistic and largely misleading to think that a state's power could be measured on the basis of the quantity and quality of military capabilities. Noting in his study that neither the United States nor the Soviet Union was able to impose its will on incomparably weaker enemies in Vietnam and Afghanistan, respectively, Luard characterized this development as 'the erosion of military power'. In particular, he contended that a great deal of states' military capabilities is either unusable or unused. When great powers use coercive force abroad, they employ only a small part of their military arsenal, even at the price of defeat. According to Luard, the main source of the erosion of military power lies in the very nature of most contemporary conflicts, which are no longer old-fashioned armed confrontations between conventional enemies, but civil wars in which major powers intervene for essentially political reasons, especially 'the type of government that holds power in other states' (Luard 1988: 11). In a world characterized by this type of conflicts, which Luard (1988: 14) described as a 'system of international civil war', the conventional arsenal of major powers is often of little use.

More than thirty years after *The Blunted Sword* was published, the trend identified by Luard seems to have strengthened. Today, most armed conflicts take place *within* states rather than between them (Pettersson et al. 2019). More importantly for this research, we are dealing with an increasing number of civil wars in which external powers intervene in support of one of the warring factions (internationalized intrastate conflicts, to employ the current jargon), as was the case in Iraq, Afghanistan, Libya, and Syria. Although, with the end of the cold war, ideology lost its primary importance, the objective of most western military interventions abroad is still the domestic situation in another

[4] Edward Luttwak (1994) famously claimed that increasing sensitivity to casualties has convinced great powers to stay at 'home with the kids'. However, the post-cold-war era attests that great powers and even medium powers are not shy of using coercive force abroad. As I clarified in Chapter 4, the prospect of casualties has affected more how force is employed than the choice of using force in the first place.

country, especially the government.[5] Virtually all conflicts in which western powers are engaged are conflicts of this sort.

Several names have been coined to describe the type of warfare typical of these armed confrontations: 'low intensity conflicts' (Creveld 1991: 58), 'new wars' (Münkler 2005; Kaldor 2012;, 'hybrid wars' (Hoffman 2007), 'post-modern wars' (Gray 1997), 'degenerate wars' (Shaw 2003), 'surrogate wars' (Krieg and Rickli 2019), and so on. But, regardless of the label employed and whatever the significant differences existing among these different types of warfare, they do not correspond to the old-fashioned clash between regular armies. In these armed confrontations, not only can a mere fraction of the military capabilities of major powers be used, or indeed has been used, but more importantly their outcome is not primarily determined by the balance of military power. One may state the obvious by remarking that, if sheer capabilities were still the decisive factor, the United States, NATO, and the other ISAF members would have been easily and swiftly victorious in Afghanistan against the Taliban and the other relatively small insurgent groups.[6] Yet, the final outcome of this and other similar conflicts is the opposite of what the military balance of the actors involved would suggest.

In most internationalized intrastate conflicts, war is not primarily fought conventionally between uniformed soldiers who try to prevail over their opponents on a well-defined and isolated battlefield. Here warfare is often geographically pervasive, with no exact localization, and at least one of the parties includes non-uniformed fighters employing the 'strategy of the weak' (Arreguín-Toft 2005), based on 'hit and run' tactics, terrorist action, guerrilla, and asymmetric warfare. These porous wars no longer have proper battlefields, friendly civilians and enemy combatants are mixed together, and the traditional mastery of the fighting ground has become largely pointless. In this kind of conflict, most heavy weaponry, as Luard noted, becomes useless. Tanks, bombers, and cruise missiles can certainly destroy buildings, villages, and entire districts, but they have proved useless in catching insurgents in densely inhabited cities or scarcely populated mountainous regions. For, being incapable of destroying the enemy army and knowing that the modern battlefield is an extremely deadly place, insurgents rarely fight in the open and prefer to employ subterranean tactics, avoiding pitched battles and direct confrontation.[7]

[5] This political goal must not be conflated with regime change, the latter being a subcategory of the larger objective mentioned here.

[6] Moreover, it must be pointed out that the wars in Iraq and Afghanistan have cost the United States around $6.4trn (Crawford 2019).

[7] There are exceptions to this rule, such as when the Taliban openly challenged NATO forces in Pashmul in September 2006. But the result, quite obviously, was a serious military defeat for the Taliban,

Even when the conventional phase of a conflict comes to an end, war may not be over, since the first regular stage—as in Iraq—might be followed by a messy one of unconventional conflict.

By denying western firepower a clear target, irregular combatants greatly reduce the destructive potential and utility of advanced military technology. Air power and high technology are devastating against low-tech enemies that fight conventionally, but they are far from being decisive when wars evolve into irregular armed confrontations. In addition, as a variety of scholars and policymakers have noted (e.g., Kissinger 1969: 214; Mack 1975: 178; Creveld 2006: 227–8), while irregular combatants win if they do not lose, conventional powers lose if they do not win. Indeed, although insurgents around the world do not entirely embrace Mao's doctrine of *protracted war*,[8] they try to prolong conflicts until they exhaust the political will of major powers.

Without buying the rhetoric of population-centric COIN as 'armed social work' (Kilcullen 2006) or suggesting the simplistic monocausal argument that the structure of counterinsurgent forces is the only variable that matters,[9] one may argue that winning an irregular conflict requires implementation of what can be described as a 'telluric' strategy, designed to establish effective control over the territory and the local population. Indeed, short of brutal barbarism in which firepower is massively and indiscriminately employed[10]—an unthinkable policy today not only for democracies (Gat 2011) but for most states—succeeding in irregular warfare requires that one employ (not only deploy) ground forces. The old flesh-body of the soldier rather than high-tech weaponry is crucial in this type of conflict. In short, the importance of the foot soldier, which diminished significantly for about a century owing to the increasing mechanization of warfare, appears to be fully restored. Yet, such an approach inevitably means exposing one's soldiers to the perils of war, a rather ordinary phenomenon in war that we are now inclined to forget or deny.

which persuaded them to avoid direct confrontation with western firepower and to go back to their usual insurgent tactics (Giustozzi 2007: 123–126).

[8] On this point, David Kilcullen (2009: 52) was correct when he argued that in Afghanistan the US was facing not a Maoist strategy, but rather an 'exhaustion' strategy meant to sap 'the energy, resources, and support of the Afghan government and its international partners'. For an analysis of Mao's doctrine of people's war, see Castelli et al. (2021). On other limits of COIN, see Zambernardi (2010).

[9] For a recent critique of the force structure thesis, based on the dichotomy between manpower and mechanization, see Scarinzi (2021).

[10] Although for western states a policy of barbarism seems unthinkable today, it must be remarked that they conducted ruthless counterinsurgency campaigns even in the post-Second World War period, with the systematic use of torture, bombing of entire villages, and chemicals to defoliate the jungle. In many of these cases, not only did western states violate basic moral principles and fundamental norms of *jus in bello*, but they did not even achieve military success.

A Sad but Simple Truth: Dying in War

'The characteristic act of men at war is not dying, it is killing,' writes historian Johanna Bourke (1999: p. xiii) in her essential study of face-to-face killing in the twentieth century. In similar vein, military historian John Keegan (1999a: 72) defined war as 'collective killing for some collective purpose'. *Pace* these one-sided definitions, war is more than killing: it is also *collective dying for a collective purpose*. War is not only about adopting the best strategy and tactics or skilful organization of logistics, but also about being able to endure tragic loss of human life. Not only is dying the inevitable other side of the combat coin, but the capacity to bear casualties has often proved a key to triumph in the bloody business of war. Even in major conventional conflicts such as the American Civil War and the First and Second World Wars, the victorious powers were also the countries that sacrificed more combatants. On the importance of dying in war, historian Martin van Creveld (2006: 228) wrote that 'compared with the willingness or the lack of it, in men (and women) to die for their cause, virtually all questions of policy, organization, doctrine, training, and equipment pale into insignificance'.[11] This is especially true when a state faces enemies that care little about their own lives, whatever the religious, ethnic, or political reasons behind this attitude. In such a context, not only must western soldiers be willing to face death in battle, but their societies, too, must be ready to sacrifice their fellow citizens if they wish to make war a useful instrument of foreign policy, whether for defending their country's national interest or for the protection of universal values and human rights abroad.

There is no doubt that inflicting casualties on western troops has become a favourite policy of insurgents, a doctrine that can be described as a coercive strategy meant to break the will of states and convince them to withdraw.[12] Sensitivity to military casualties encourages enemies to resist, because they know that time and western deaths are on their side. Whether the use of force can succeed in achieving the ends of politics partly (though not solely, of course) depends on the political will of states to sacrifice the lives of their combatants. That may sound rather cynical and insensitive from the soldiers' viewpoint, but war does not come cheap. In Peter Paret's commentary (1986: 200–1) on Clausewitz's philosophy of war,

[11] According to Herodotus, 'a wish to kill but not to die' is a very dangerous disposition in war (in Hanson 1989: 10). Herodotus attributed such an inclination to the Persians. That seems to be the case both in ancient times and in the present era, and it is especially true in irregular warfare. Although soldiers are trained to face death in battle, the main problem appears to lie at the societal level.

[12] For a similar argument employed to explain suicide terrorism, see Pape (2005).

> the armed forces do not exist for their own sake. They are an instrument to be used ... If the political purpose demands it, the armed forces must be content with the partial mobilization of resources and with limited achievements; or, on the other hand, *they must be prepared to sacrifice themselves, and neither society nor government should regard this sacrifice, if it is an expression of rational policy, as beyond their mission.* (emphasis added)

Although casualty minimization is certainly a domestic imperative for the continuation of military intervention, it seems to be turning into a self-defeating policy as a response to particular international threats. The fact is that, while the West is willing to spend enormous financial resources on military equipment and technology, it is not ready to sacrifice the lives of its men and women. Of course, risking life is no recipe for victory. States can hazard and expend the life of their soldiers in a variety of ways; and in many instances they have done so in foolish, useless, and counterproductive ways. But what the West does not accept is the inevitable human cost of war, a cost that can be reduced but not cancelled out even by the most advanced technology. Technological improvements have provided the West with tremendous military advantages and with unparalleled strategic freedom, but fighting in a casualty-averse way against opponents who hold their own lives cheap poses a dilemma that the high-tech approach to war has not yet solved. For, once unprecedented importance has been attached to human losses, the advantages resulting from the combination of wealth and technology greatly diminish.

The indecisive results of the western way of war in recent conflicts bear limited but significant witness to the fact that superior numbers and superior technology are no use if they are not backed up by firm popular support for the inevitable human cost that war implies. In other words, whereas Luard (1988) noted that much of the military arsenal of major powers is unusable in today's armed conflicts, what has happened with the increasing sensitivity to military casualties is that another essential means of war (that is, soldiers) has become largely unusable.

But can states change their way of war? No doubt, when a particular type of warfare no longer achieves the expected results, a strong pressure for change usually develops. Especially after burning and unexpected defeats, modifications in the conduct of war are likely to occur. In particular, as military historian John Lynn (2003: 360) contended, when a military approach fails, there is often a feedback process inducing a state to reconsider its strategy and tactics. On the possibility of modifying the American way of war in the aftermath of the Kosovo conflict, Eliot Cohen (2001: 59) maintained that 'barring some cataclysmic event—a twenty-first-century Pearl Harbor—it seems likely that the American way of war will prevail for some time to come'. These words were written on the eve of 11 September 2001, in its own right a sort of cataclysmic event. The attacks on the heart of the United States seemed to reactivate the traditional patriotism and self-sacrifice of American society. Not only did many citizens rally round the flag, but many also decided that the lives of their fellow uniformed citizens could be sacrificed for the security of the nation. Yet, even though Americans and Europeans were ready to sacrifice a few thousand soldiers in Iraq and Afghanistan, we know that they eventually decided to forfeit the political goals pursued in these conflicts rather than sustain further casualties. Colin Gray (2006: 18) appears to have been right when he pointed out that, even when a state can recognize some significant malfunctions in its strategic and military cultures, it 'may not be able to take effective corrective action'. A military lesson may be learned, but with no power to put it into practice.

To Conclude: Political Ends without Military Means

In his *magnum opus*, Carl von Clausewitz (1984: 87) famously declared that 'war is not merely an act of policy but a true political instrument, a continuation of political intercourse, carried on with other means'.[13] According to the Prussian general, war should be conceived and conducted as an instrument of state policy, not as an independent, self-sufficient activity, since its rationale depends on the objectives set by the political leadership. In commenting on Clausewitz's dictum, Peter Paret (1986: 200) explained that 'violence should express the political purpose, and express it in a rational, utilitarian manner; it should not take the place of the political purpose, nor obliterate it'. Thus, Clausewitz's formula implies a strong prescriptive component: the primacy of politics over the military management of war. Although Clausewitz recognized that violence tends to extremes in that its dynamics may develop independently and endanger political rationality, he believed that it was the task precisely of the political leadership to employ war for the ends of politics. In

[13] Given the radical changes that war and its conduct have brought about in the twentieth and twenty-first centuries, a few prominent students of war, notably Martin van Creveld (1991: 49–62), John Keegan (1993: 3–23), and Mary Kaldor (2012: 17), have questioned the relevance of Clausewitz to an appreciation of war in the contemporary world. Although some parts of his *On War* are certainly interesting for military historians alone (e.g., Clausewitz's writings provide no analysis of the role of economics, technology, or naval power in warfare, see Moran 2007: 92 and Heuser 2007, pp. xxxi-xxxii), his general philosophy and especially his grasp of the relationship between military means and political ends still remain crucial today.

other words, military strategy and tactics should be devised according to these ends, which, in turn, should determine the choice of the means employed and the conduct of operations.

Typically, twentieth-century militarism came to overthrow the supremacy of politics over the conduct of war. The core of militarism was expressed with unequivocal clarity by German general Erich Ludendorff, who, reversing Clausewitz's insight, wrote that 'politics subserves the conduct of war … Sound policy is the continuation of war by other means in peacetime' (in Ritter 1969: 6). As an exaggeration and prioritization of the military factors, militarism corrupts war as a tool of politics. Whereas militarism undoubtedly undermines the controlling effects of politics on war, it is nevertheless far from being the only way in which the Clausewitzian primacy of politics over military means may be reversed. It is a major implication of this research that the present meaning ascribed to military deaths has eroded the supremacy that politics is expected to establish over the military instrument and, as a result, is jeopardizing the use of war as a rational instrument.

The conduct of war and the choice of means to employ must be conditioned by the objectives of state policy. That is Clausewitz's timeless lesson. By contrast, casualty-averse warfare—intended as a broad set of military policies aiming at minimizing friendly casualties—is an approach that is being adopted regardless of war's political purpose. It is the primary goal of this doctrine to prevent or at least minimize western casualties, with little concern for the consequences on the political objectives of war. From a strategic viewpoint, the choice of specific means should be related to the political end at issue. However, means are now being chosen at the expense of national-security goals. While force protection measures, air power, the increasing use of unmanned aerial venhicles (UAVs), and other military policies meant to reduce friendly casualties show a noble humanitarian concern for western soldiers' lives, they are unsuitable for the type of irregular warfare in which western powers are engaged. A preference for employing machines rather than soldiers can help with the home front and with a domestic public opinion grown increasingly sensitive to casualties (Bacevich 2013: 25), but not with the political aims prompting the use of force, because such a preference allows soldiers' lives to outweigh political ends and military means to prevail over the political rationale of war. Especially in irregular conflicts, warfare driven by casualty sensitivity breaks the relationship between politics and military means, thus jeopardizing the political effectiveness of the use of force.[14]

[14] Clausewitz devoted many writings to irregular wars and warfare—what he referred to as small wars. See Heuser (2010), Daase and Davis (2015), and Scheipers (2018).

As irregular wars are now playing a predominant role in the international scenario, westerners are likely to be engaged quite often in conflicts that they are not prepared to fight.[15] In fact, it is not coincidental, but logical, that the military superiority of the West has been countered by unconventional strategies and tactics on the part of enemies that would be crushed if they fought on western terms. In a world in which military conflicts are mostly fought 'amongst' and, especially, 'for the people' (Smith 2006), victory will be secured not by the development of new high-tech weapons and novel military strategies but rather by the political ability to accept and tolerate the human cost of war.

Arguing that the value that is attached to our soldiers' lives has led to a further erosion of western military power does not imply that we should simply disregard our human losses. Reversing the priority of protecting our soldiers' lives appears to be unthinkable today. The primary importance attached to the lives of soldiers is a humane development that deserves admiration.[16] Actually, one may note that too many lives have been needlessly expended in recent armed conflicts. What we need to admit are the social limits of current western military power, because it is no longer reasonable to start wars whose human costs we cannot sustain.

On this point, a final Clausewitzian note is in order. Just before formulating his famous dictum on 'war as a continuation of politics with other means', the Prussian general wrote that, although the political goal is 'the supreme consideration in conducting' war (Clausewitz 1984: 87), it must still adapt to the nature of the means available. We may conclude by suggesting that, if the nature of military means does not allow us to achieve the ends of politics, instruments other than war must be employed.

[15] One may rightly argue that casualty aversion would virtually disappear if national survival were at stake in so-called wars of necessity, where an existential threat to a society is present. For a (self-) critique of the distinction between 'wars of necessity' and 'wars of choice', see however Freedman (2010).

[16] Likewise, the attempt to make war less cruel is a commendable effort. However, as Samuel Moyn (2021) has recently argued, the US attempt to humanize war has had the effect of sanitizing the use of force and producing 'forever war'.

References

Aberbach, D. (1996). *Charisma in Politics, Religion and the Media: Private Trauma, Public Ideals*. London: Palgrave Macmillan.

Adamson, P. B. (1977). 'A Comparison of Ancient and Modern Weapons in the Effectiveness of Producing Battle Casualties', *Journal of the Royal Army Medical Corps*, 123: 93–103.

Agamben, G. (1998). *Homo Sacer: Sovereign Power and Bare Life*. Stanford: Stanford University Press.

Allen, M. J. (2009). *Until the Last Man Comes Home: POWs, MIAs, and the Unending Vietnam War*. Chapel Hill: University of North Carolina Press.

Althaus, S. L., B. H. Bramlett, and J. G. Gimpel (2012). 'When War Hits Home: The Geography of Military Losses and Support for War in Time and Space', *Journal of Conflict Resolution*, 56/3: 382–412.

Anders, S. E. (1988). 'With All Due Honors: A History of the Quartermaster Graves Registration Mission', *Quartermaster Professional Bulletin* (September), https://www.quartermasterfoundation.org/article/with-all-due-honors-a-history-of-the-quartermaster-graves-registration-mission/.

Anderson, B. (1991). *Imagined Communities: Reflections on the Origin and Spread of Nationalism*. New York: Verso.

Appiah, K. A. (2016). 'There Is No Such Thing as Western Civilization', *Guardian*, 9 November, https://www.theguardian.com/world/2016/nov/09/western-civilisation-appiah-reith-lecture.

Arias, E. (2019). 'United States Life Tables, 2017', *National Vital Statistics Reports*, 68/7: 1–65.

Ariès, P. (1981). *The Hour of our Death*. New York: Alfred A. Knopf.

Ariès, P. (1985). *Images of Man and Death*. Cambridge, MA: Harvard University Press.

Ariès, P. (1996). Centuries of Childhood. London: Pimlico.

Ariès, P. (2009). *Western Attitudes toward Death: From the Middle Ages to the Present*. Baltimore: Johns Hopkins University Press.

Aristotle (1998). *Nicomachean Ethics: Book VIII and IX*, ed. M. Pakaluk. Oxford: Oxford University Press.

Arnold, T. F. (1999). 'War in the Sixteenth-Century Europe: Revolution and Renaissance', in J. Black (ed.), *European Warfare 1453–1815*. London: Macmillan, 23–44.

Aron, R. (1983). *Clausewitz: Philosopher of War*. London: Routledge & Kegan Paul.

Arreguín-Toft, I. (2005). *How the Weak Win Wars: A Theory of Asymmetric Conflict*. New York: Cambridge University Press.

Arrington, N. T. (2014). *Ashes, Images and Memories: The Presence of the War Dead in Fifth Century Athens*. Oxford: Oxford University Press.

Asch, R. G. (1997). *The Thirty Years' War: The Holy Roman Empire and Europe 1618–48*. New York: Macmillan.

Åse, C., and M. Wendt (2017). 'Gendering the New Hero Narratives: Military Death in Denmark and Sweden', *Cooperation and Conflict*, 53/1: 23–41.

Auchter, J. (2015). 'Corpses', in M. B. Salter (ed.), *Making Things International 1: Circuits and Motion*. Minneapolis: University of Minnesota Press, 129–40.

Audoin-Rouzeau, S. (1996). 'The French Soldier in the Trenches', in H. Cecil and P. H. Liddle (eds), *Facing Armageddon: The First World War Experienced*. London: Leo Cooper, 221–9.

Bacevich, A. J. (2013). *Breach of Trust: How Americans Failed their Soldiers and their Country*. New York: Metropolitan Books.

Bacevich, A. J., and E. A. Cohen (2001). 'Introduction: Strange Little War', in A. J. Bacevich and E. A. Cohen (eds), *War over Kosovo: Politics and Strategy in a Global Age*. New York: Columbia University Press, pp. ix–xiv.

Bachleitner, K. (2021). *Collective Memory in International Relations*. Oxford: Oxford University Press.

Barber, B. (1949). 'Place, Symbol and Utilitarian Function in War Memorials', *Social Forces*, 28/1: 64–8.

Barrett, M. (2014). '"White Graves": The Imperial War Graves Commission in East and West Africa, 1918–1939', in P. Cornish and N. J. Saunders (eds), *Bodies in Conflict: Corporeality, Materiality and Transformation*. London and New York: Routledge, 80–90.

Bartov, O. (1992). *Hitler's Army: Soldiers, Nazis, and War in the Third Reich*. New York and Oxford: Oxford University Press.

Baruch, M. C., and E. J. Beckman (1978). *Civil War Union Monuments: A List of Union Monuments, Markers and Memorials of the American Civil War, 1861–1865*. Washington: Daughters of Union Veterans of the Civil War.

Batkin, L. M. (1992). *L'idea di individualità nel Rinascimento italiano*. Rome and Bari: Laterza.

Beauvoir, S. de (1966). *A Very Easy Death*. New York: G. P. Putnam's Sons.

Beck, U. (1992). *Risk Society: Towards a New Modernity*. London: Sage.

Becker, A. (1989). *Les Monuments aux morts*. Paris: Errance.

Becker, A. (1992). 'From Death to Memory: National Ossuaries in France after the Great War', *History and Memory*, 5/2: 32–49.

Becker, A. (1997). 'War Memorials: A Legacy of Total War?', in S. Förster and J. Nagler (eds), *On the Road to Total War: The American Civil War and the German Wars of Unification, 1861–1871*. New York: Cambridge University Press, 657–80.

Becker, E. (1973). *The Denial of Death*. New York: Free Press.

Beehner, L. (2006). 'US Military Strategy in Iraq', *Council on Foreign Relations*, 12 April, http://www.cfr.org/publication/10434/us_military_strategy_in_iraq.html.

Bellamy, A. J. (2006). *Just Wars: From Cicero to Iraq*. Cambridge: Polity.

Ben-Amos, A. (2000). *Funerals, Politics, and Memory in Modern France*. Oxford: Oxford University Press.

Ben-Ari, E. (2005). 'Epilogue: A "Good" Military Death', *Armed Forces & Societies*, 31/4: 651–64.

Benhabib, S. (2006). 'The Philosophical Foundations of Cosmopolitan Justice', in R. Post (ed.), *Another Cosmopolitanism*. Oxford and New York: Oxford University Press, 13–36.

Berinsky, A. J. (2009). *In Time of War: Understanding American Public Opinion from World War II to Iraq*. Chicago: University of Chicago Press.

Best, G. (1998). *War and Society in Revolutionary Europe, 1770–1870*. Stroud: Sutton.

Biddle, S. (2004). *Military Power: Explaining Victory and Defeat in Modern Battle*. Princeton: Princeton University Press.

Biddle S., J. A. Friedman, and J. N Shapiro (2012). 'Testing the Surge: Why Did Violence Decline in Iraq in 2007?', *International Security*, 37/1: 7–40.

Biden, J. (2021). 'Remarks by President Biden on Afghanistan', The White House, 12 August, https://www.whitehouse.gov/briefing-room/speeches-remarks/2021/08/16/remarks-by-president-biden-on-afghanistan/.

Bingham, P. M. (1999). 'Human Uniqueness: A General Theory', *Quarterly Review of Biology*, 74/2: 133–69.

Biow, D. (2015). *On the Importance of Being an Individual in Renaissance Italy*. Philadelphia: University of Pennsylvania Press.

Black, J. (1999). 'Introduction', in J. Black (ed.), *European Warfare 1453–1815*. London: Macmillan, 1–22.

Black, J. (2005). *Introduction to Global Military History: 1775 to the Present Day*. Abingdon and New York: Routledge.

Black, M. (2010). *Death in Berlin: From Weimar to Divided Germany*. New York: Cambridge University Press.

Blair, W. A. (2004). *Cities of the Dead: Contesting the Memory of the Civil War in the South, 1865–1914*. Chapel Hill: University of North Carolina Press.

Blanco, R. L. (1965). 'Reform and Wellington's Post Waterloo Army, 1815–1854', *Military Affairs*, 29/3: 123–32.

Bloch, M., and J. P. Parry (1982) (eds). *Death and the Regeneration of Life*. Cambridge: Cambridge University Press.

Blunden, E. (2003). 'Introduction', in P. Longworth (ed.), *The Unending Vigil: The History of the Commonwealth War Graves Commission*. Barnsley: Leo Cooper, pp. xvii–xxv.

Boettcher, W. A., and M. D. Cobb (2009). 'Don't Let Them Die in Vain: Casualty Frames and Public Tolerance for Escalating Commitment in Iraq', *Journal of Conflict Resolution*, 53/5: 677–97.

Bogacz, T. (1986). '"A Tyranny of Words": Language, Poetry, and Antimodernism in England in the First World War', *Journal of Military History*, 58/3: 643–68.

Bogorov, V. (2002). 'In the Temple of Sacred Motherland: Representations of National Identity in the Soviet and Russian WWII Memorials', Department of Geography, University of Wisconsin-Madison, http://www.dartmouth.edu/~crn/groups/geographies_group_papers/Finalpapers/Bogorov02.pdf.

Bollet, A. J. (2002). *Civil War Medicine: Challenges and Triumphs*. Tucson, AZ: Galen Press.

Boorman, D. (1995). *For your Tomorrow: British Second World War Memorials*. York: Ebor Press.

Borg, A. (1991). *War Memorials: From Antiquity to the Present*. London: Leo Cooper.

Bourke, J. (1999). *An Intimate History of Killing: Face to Face Killing in the 20th Century Warfare*. New York: Basic Books.

Bourne, J. (2005). 'Total War I: The Great War', in C. Townshend (ed.), *The Oxford History of Modern War*. New York: Oxford University Press, 117–37.

Bowker, J. (1991). *The Meanings of Death*. Cambridge: Cambridge University Press.

Boyle, M. J. (2020). *The Drone Age: How Drone Technology Will Change War and Peace*. Oxford: Oxford University Press.

'Brady's Photographs: Pictures of the Dead at Antietam' (1862). *New York Times* (October), 5, https://www.nytimes.com/1862/10/20/archives/bradys-photographs-pictures-of-the-dead-at-antietam.html.

Brandt, S. (1994). 'The Memory Makers: Museums and Exhibitions of the First World War', *History and Memory*, 6/1: 95–122.

Braudel, F. (1958). 'Histoire et sciences sociales: La Longue Durée', *Annales: Économies, Sociétés, Civilisations*, 13/4: 725–53.

Brearton, F. (2003). *The Great War in Irish Poetry*. Oxford and New York: Oxford University Press.

Bregantin, L. (2010). *Per non morire mai: La percezione della morte in guerra e il culto dei caduti nel primo conflitto mondiale*. Padua: Il Poligrafo.

Bregantin, L., and B. Brienza (2015). *La guerra dopo la guerra: Sistemazione e tutela della salme dei caduti della Grande Guerra dai cimiteri al fronte ai sacrari monumentali*. Padua: Il Poligrafo.

British Academy (2012). *Cultures of Commemoration: War Memorials, Ancient and Modern*. Oxford: Oxford University Press.

Brown, P. (2003). *The Rise of Western Christendom*. Oxford: Blackwell.

Brown, V. (2008). *The Reaper's Garden: Death and Power in the World of Atlantic Slavery*. Cambridge, MA: Harvard University Press.

Browning, P. (2002). *The Changing Nature of Warfare: The Development of Land Warfare from 1792 to 1945*. Cambridge: Cambridge University Press.

Budreau, L. M. (2010). *Bodies of War: World War I and the Politics of Commemoration in America, 1919–1933*. New York and London: New York University Press.

Buffton, D. D. (2005). 'Memorialization and the Selling of War', *Peace Review: A Journal of Social Justice*, 17/1: 25–31.

Bugnion, F. (2009). 'From Solferino to the Birth of Contemporary International Humanitarian Law', 22 April, https://www.icrc.org/en/doc/assets/files/other/solferino-bugnion-icrc.pdf.

Buley, B. (2008). *The New American Way of War: Military Culture and the Political Utility of Force*. Routledge: London.

Burckhardt, J. (1944). *The Civilization of the Renaissance in Italy*. Vienna: Phaidon.

Burk, J. (1999.) 'Public Support for Peacekeeping in Lebanon and Somalia: Assessing the Casualties Hypothesis', *Political Science Quarterly*, 114/1: 53–78.

Burton, R. D. E. (2001). *Blood in the City: Violence and Revelation in Paris, 1789–1945*. Ithaca, NY, and London: Cornell University Press.

Bushaway, B. (1992). 'Name upon Name: The Great War and Remembrance', in R. Porter (ed.), *Myths of the English*. Cambridge: Polity, 136–67.

Bushaway, B. (2001). 'The Obligation of Remembrance or Remembrance of Obligation: Society and the Memory World War', in J. Bourne, P. Liddle, and I. Whitehead (eds), *The Great World War 1919–45, Volume 2: The People's Experience*. London: Harper Collins, 491–507.

Butler, J. (2004). *Precarious Life: The Powers of Mourning and Violence*. London: Verso.

Butler, J. (2009). *Frames of War: When is Life Grievable?* London: Verso.

Canella, M. (2010). *Paesaggi della morte: Riti, sepolture e luoghi funerari tra Settecento e Novecento*, with a preface of Michel Vovelle. Rome: Carocci.

Caniglia, R. (2001). 'US and British Approaches to Force Protection', *Military Review*, 81/4: 73–81.

Cannadine, D. (1981). 'War and Death, Grief and Mourning in Modern Britain', in J. Whaley (ed.), *Mirrors of Mortality: Studies in the Social History of Death*. New York: St Martin's Press, 187–242.

Capdevila, L., and D. Voldman (2006). *War Dead: Western Societies and the Casualties of War*. Edinburgh: Edinburgh University Press.

Cardini, F. (2013). *Quell'antica festa crudele: Guerra e cultura della guerra dal Medioevo alla Rivoluzione francese*. Bologna: Il Mulino.

Carlton, C. (2003). *Going to the Wars: The Experience of the British Civil Wars, 1638–1651*. London and New York: Routledge.

Carpenter, C. (2011). 'Fighting the Laws of War: Protecting Civilians in Asymmetric Conflict', *Foreign Affairs*, 90/2: 146–52.

Cassidy, B. (2003). 'Machiavelli and the Ideology of the Offensive: Gunpowder Weapons in *The Art of War*', *Journal of Military History*, 67/2: 381–404.

Cassidy, R. M. (2006). 'The Long Small War: Indigenous Forces for Counterinsurgency', *Parameters*, 36/2: 47–62.

Castelli, E., S. Dossi, and L. Zambernardi (2021). 'Lost in Transition: The Myth of Mao and the Origins of COIN', *Small Wars & Insurgencies*, 32/2: 320–42.

Castelli, E., and L. Zambernardi (2017). 'Force Protection and its Trade-offs', in D. Reiter (ed.), *The Sword's Other Edge: Trade-offs in the Pursuit of Military Effectiveness*. Cambridge: Cambridge University Press, 31–57.

Caverley, J. D. (2014). *Democratic Militarism: Voting, Wealth, and War*. Cambridge: Cambridge University Press.

Caverley, J. D. (2016). 'Roundtable 8–12 on *Democratic Militarism: Voting, Wealth, and War*: Author's Response by Jonathan D. Caverley', *H-Diplo/ISSF*, 25 April, https://issforum.org/roundtables/8-12-democratic-militarism#Authors_Response_by_Jonathan_D_Caverley_Massachusetts_Institute_of_Technology.

Chabod, F. (2008). *Machiavelli and the Renaissance*. ACLS Humanities E-Book.

Chamayou, G. (2015). *Drone Theory*. London: Penguin.

Chaunu, P. (1978). *La Mort à Paris: 16 e, 17 e, 18 e siècles*. Paris: Fayard.

Chernow, R. (2010). *Washington: A Life*. New York: Penguin.

Churchill, W. S. (2005). *The World Crisis 1911–1918*. New York: Free Press.

Clarke, J. (2007). *Commemorating the Dead in Revolutionary France: Revolution and Remembrance, 1789–1799*. Cambridge: Cambridge University Press.

Clausewitz, C. von (1984). *On War*, ed. and trans. M. Howard and P. Paret. Princeton: Princeton University Press.

Clodfelter, M. (2008). *Warfare and Armed Conflicts: A Statistical Encyclopedia of Casualty and Other Figures, 1494–2007*. Jefferson, NC: McFarland & Company.

Clodfelter, M. (2010a). *Beneficial Bombing: The Progressive Foundations of American Air, 1917–1945*. Lincoln: University of Nebraska Press.

Clodfelter, M. (2010b). 'Aiming to Break Will: America's World War II Bombing of German Morale and its Ramifications', *Journal of Strategic Studies*, 33/3: 401–35.

Coco, G. A. (1995). *A Strange and Blighted Land: Gettysburg, The Aftermath of a Battle*. Gettysburg, PA: Thomas Publications.

Cohen, E. A. (2001). 'Kosovo and the New American Way of War', in A. J. Bacevich and E. A. Cohen (eds), *War over Kosovo: Politics and Strategy in a Global Age*. New York: Columbia University Press, 38–62.

Coker, C. (2007). *The Warrior Ethos: Military Culture and the War on Terror*. London: Routledge.

Coldstream & District Local History Society, 'War Memorials, https://www.coldstreamhistorysociety.co.uk/category/projects/war-memorials/.

Colombo, A. D. (2017). 'Why Europe Has Never Been United (not Even in the Afterworld): The Fall and Rise of Cremation in Cities (1876–1939)', *Death Studies*, 41/1: 22–33.

Colombo, A. D. (2021). *La solitudine di chi resta: La morte al tempo del contagio*. Bologna: Il Mulino.

Congressional Research Services (2019). *American War and Military Operations Casualties: Lists and Statistics*, CRS Report RL32492, https://www.fas.org/sgp/crs/natsec/RL32492.pdf.

Connelly, M. (2002). *The Great War, Memory and Ritual: Commemoration in the City and East London, 1916–1939*. Woodbrige: Boydell Press.

Convention for the Amelioration of the Condition of the Wounded and Sick in Armies in the Field (1929). Geneva, 27 July, https://ihl-databases.icrc.org/ihl/INTRO/300?OpenDocument.

Conway, S. R. (1987). 'The Recruitment of Criminals into the British Army, 1775–81', *Historical Research*, 58/137: 46–58.

Corvisier, A. (1979). *Armies and Society in Europe, 1494–1789*. Bloomington: University of Indiana University Press.

Corvisier, A. (1985). *Les Hommes, la guerre et la mort*. Paris: Economica.

Coser, L. A. (1977). *Masters of Sociological Thought: Ideas in Historical and Social Context*. New York: Harcourt, Brace, Jovanovich.

Coss, E. J. (2005). 'All for the King's Shilling: An Analysis of the Campaign and Combat Experiences of the British Soldier in the Peninsular War, 1808–1914', Dissertation, Columbus: Ohio State University.

Coss, E. J. (2010). *All for the King's Shilling: The British Soldier under Wellington, 1808–1814*. Norman: University of Oklahoma Press.

Cox, M. (2014). 'Industry: Iron Man Still Hollywood, not Reality', 22 April, http://www.military.com/daily-news/2014/04/22/industry-iron-man-still-hollywood-not-reality.html?ESRC=todayinmil.sm.

Craig, G. A. (1986). 'Delbrück: The Military Historian', in P. Paret (ed.), with the collaboration of G. A. Craig and F. Gilbert, *Makers of Modern Strategy: From Machiavelli to the Nuclear Age*. Princeton: Princeton University Press, 326–53.

Crawford, N. (2019). 'United States Budgetary Costs and Obligations of Post-9/11 Wars through FY2020: $6.4 Trillion', Watson Institute for International and Public Affairs, Brown University, 13 November, https://watson.brown.edu/costsofwar/files/cow/imce/papers/2019/US%20Budgetary%20Costs%20of%20Wars%20November%202019.pdf.

Creveld, M. van (1989). *Technology and War: From 2000 BC to the Present*. New York: Free Press.

Creveld, M. van (1991). *The Transformation of War*. New York: Free Press.

Creveld, M. van (2004). *Supplying War: Logistics from Wallenstein to Patton*. Cambridge: Cambridge University Press.

Creveld, M. van (2006). *The Changing Face of War: Lessons of Combat, from the Marne to Iraq*. New York: Presidio Press.

Creveld, M. van (2008). *The Culture of War*. New York: Presidio Press.

Crossland, Z. (2000). 'Buried Lives: Forensic Archaeology and the Disappeared in Argentina', *Archaeological Dialogues*, 7/2: 146–59.

Daase, C., and J. W. Davis (2015). *Clausewitz on Small War*. Oxford: Oxford University Press.

Damousi, J. (1999). *The Labour of Loss: Mourning, Memory and Wartime Bereavement in Australia*. Cambridge: Cambridge University Press.

Davies, D. J. (2005). *A Brief History of Death*. Malden, MA: Blackwell Publishing.

De Luna, G. (2006). *Il corpo del nemico ucciso: Violenza e morte nella guerra contemporanea*. Turin: Einaudi.

Defense Prisoner of War/Missing Personnel Office. 2014. 'Summary Statistics'. http://www.dtic.mil/dpmo/summary_statistics/ .accessed 17 January 2014.

Degler, C. N. (1997). 'The American Civil War and the German Wars of Unification: The Problem of Comparison', in S. Förster and J. Nagler (eds), *On the Road to Total War:*

The American Civil War and the German Wars of Unification, 1861–1871. New York: Cambridge University Press, 53–71.
Delmas, J. (1992) (ed.). *Histoire militaire de la France, volume II: De 1715 à 1871*. Paris: Presses Universitaires de France.
Dendooven, D. (2014). '"Bringing the Dead Home": Repatriation, Illegal Repatriation and Expatriation of British Bodies during and after the First World War', in P. Cornish and N. J. Saunders (eds), *Bodies in Conflict: Corporeality, Materiality and Transformation*. London and New York: Routledge, 66–79.
Dennie, G. (2009). 'The Standard of Dying: Race, Indigence, and the Disposal of the Dead Body in Johannesburg, 1886–1990', *African Studies*, 68/3: 310–30.
Der Derian, J. (2009). *Virtuous War: Mapping the Military–Industrial–Media–Entertainment Network*. New York: Routledge.
Dogliani, P. (1986). 'La memoria della Grande Guerra in Francia nel primo e secondo dopoguerra', in D. Leoni and C. Zadra (eds), *La Grande Guerra: Esperienza, memoria, immagini*. Bologna: Il Mulino, 545–74.
Dogliani, P. (1995). 'Luogi della memoria e monumenti', in B. Dalla Casa and A. Preti (eds), *Bologna in guerra 1940–1945*. Milan: FrancoAngeli, 461–75.
Dogliani, P. (1996). 'Redipuglia', in M. Isnenghi (ed.), *I luoghi della memoria: Simboli e miti dell'Italia unita*. Bari and Rome: Laterza, 375–89.
Dogliani, P. (1999). 'Constructing Memory and Anti-Memory: The Monumental Representation of Fascism and its Denial in Republican Italy', in R. J. B. Bosworth and P. Dogliani (eds), *Italian Fascism: History, Memory, and Representation*. London: Palgrave Macmillan, 11–30.
Dogliani, P. (2006). 'I monumenti e le lapidi come fonti', in C. Pavone (ed.), *Storia d'Italia nel secolo ventesimo. Strumenti e fonti, Vol. 2. Istituti, musei e monumenti, bibliografia e periodici, associazioni, finanziamentei per la ricerca*. Rome: Ministero per i beni e le attività culturali, 261–75.
Dogliani, P. (2009) 'Il cimitero tedesco della Futa', in M. Maggiorani and P. Zagatti (eds), *La montagna dopo la guerra. Continuità e rotture nell'Appennino bolognese tra Idice e Setta-Reno: 1945-2000*. Bologna: Aspasia, 475–85.
Doss, E. (2008). 'War, Memory, and the Public Mediation of Affect: The National World War II Memorial and American Imperialism', *Memory Studies*, 1/2: 227–50.
Doty, R. L. (2011). 'Bare Life: Border-Crossing Deaths and Spaces of Moral Alibi', *Environmental and Planning D: Society and Space*, 29/4: 599–612.
Douglas, R. (2021). 'TALOS Project: How the US Military Tried to Make an "Iron Man" Suit', *National Interest*, 10 March, https://nationalinterest.org/blog/buzz/talos-project-how-us-military-tried-make-iron-man-suit-179863.
Dower, J. W. (1986). *War without Mercy: Race and Power in the Pacific War*. New York: Pantheon.
Downes, A. B. (2008). *Targeting Civilians in War*. Ithaca, NY: Cornell University Press.
Drévillon, H. (2013). *L'Individu et la guerre: Du chevalier Bayard au soldat inconnu*. Paris: Belin.
Dumont, L. (1992). *Essays on Individualism: Modern Ideology in Anthropological Perspective*. Chicago: University of Chicago Press.
Dupuy, R. E., and T. N. Dupuy (1956). *Military Heritage of America*. New York: McGraw-Hill.
Dupuy, T. N. (1984). *The Evolution of Weapons and Warfare*. New York: Da Capo Press.
Durkheim, E. (2002). *Suicide: A Study in Sociology*. London: Routledge.

Echevarria, A. J., II (2007). *Clausewitz and Contemporary War*. Oxford: Oxford University Press.

Egenes, K., and F. Vlasses (2016). 'Clara Barton: Angel of the Battlefield', in D. A. Forrester (ed.), *Nursing's Greatest Leaders: A History of Activism*. New York: Springer, 71–104.

Ehrenreich, B. (1997). *Blood Rites: Origins and History of the Passion of War*. New York: Owl Books.

Elias, N. (2000). *The Civilizing Process: Sociogenetic and Psychogenetic Investigations*. Malden, MA: Blackwell.

Elias, N. (2001a). *The Society of Individuals*. New York and London: Continuum.

Elias, N. (2001b). *The Loneliness of the Dying*. New York and London: Continuum.

Elliot, H. W. (1999). 'Identification', in R. Gutman and D. Rieff (eds), *Crimes of War: What the Public Should Know*. New York and London: W. W. Norton & Co., 232–3.

Ellis, J. (1986). *The Social History of the Machine Gun*. Baltimore: Johns Hopkins Uni)versity Press.

Erichsen, H. (1887). *The Cremation of the Dead*. Detroit: D. O. Haynes & Company.

Erickson, J. (1983). *The Road to Berlin: Stalin's War with Germany, Volume II*. London: Weidenfeld and Nicolson.

Etlin, R. A. (1984). *The Architecture of Death: The Transformation of the Cemetery in Eighteenth-Century Paris*. Cambridge, MA: MIT Press.

Evera, S. van (1984). 'The Cult of the Offensive and the Origins of the First World War', *International Security*, 9/1: 58–107.

Everts, P., and P. Isernia (2015). *Public Opinion, Transatlantic Relations and the Use of Force*. New York: Palgrave Macmillan.

Fallaci, O. (1976). *Interview with History*. New York: Liveright.

Farrell, J. J. (1980). *Inventing the American Way of Death, 1830–1920*. Philadelphia: Temple University Press.

Farrell, T. (2005). *The Norms of War: Cultural Beliefs and Modern Conflict*. Boulder, CO: Lynne Rienner.

Farrell, T. (2018). *Unwinnable: Britain's War in Afghanistan, 2001–2014*. London: Vintage.

Faust, D. G. (2008). *This Republic of Suffering: Death and the American Civil War*. New York: Alfred A. Knopf.

Favole, A. (2003). *Resti di umanità. Vita sociale del corpo dopo la morte*. Rome and Bari: Laterza.

Fazal, T. M. (2014). 'Dead Wrong? Battle Deaths, Military Medicine, and Exaggerated Reports of War's Demise', *International Security*, 39/1: 95–125.

Feaver, P. D. (2003). *Armed Servants: Agency, Oversight, and Civil–Military Relations*. Cambridge, MA: Harvard University Press.

Feaver, P. D., and C. Miller (2014). 'Provocations on Policymakers: Casualty Aversion and Post-Heroic Warfare', in S. Scheipers (ed.),, *Heroism and the Changing Character of War: Toward Post-Heroic Warfare*. New York: Palgrave Macmillan, 145–61.

Fehrenbach, T. R. (1994). *This Kind of War: The Classic Korean War History*. Washington: Brassey's.

Ferguson, N. (1998). *The Pity of War*. London: Allen Lane.

Ferrándiz, F., and A. C. G.M. Robben (2015) (eds). *Necropolitics: Mass Graves and Exhumation in the Age of Human Rights*. Philadelphia: University of Pennsylvania Press.

Figes, O. (2010). *Crimea*. London: Allen Lane.

Foltyn, J. (2008). 'The Corpse in Contemporary Culture: Identifying, Transacting, and Recoding the Dead Body in the Twenty-First Century', *Mortality*, 13/2: 99–104.

Foucault, M. (2003). *"Society Must Be Defended": Lectures at the Collège de France, 1975–76.* New York: Picador.

Fowler, B. (2005). 'Collective Memory and Forgetting: Components for a Study of Obituaries', *Theory, Culture & Society*, 22/6: 53–72.

Fowler B., and E. Bielsa (2007). 'The Lives We Choose to Remember: A Quantitative Analysis of Newspaper Obituaries', *Sociological Review*, 55/2: 203–26.

Fredman, A., and G. Steinberg (2010). 'Precision-Guided or Indiscriminate? NGO Reporting on Compliance with the Laws of Armed Conflict', *NGO Monitor*, 28 June, Jerusalem.

Freedman, L. (2010). 'On War and Choice', *National Interest*, 107 (May–June), 9–16.

French, D. (2005). 'The Nation in Arms II: The Nineteenth Century', in C. Townshend (ed.), *The Oxford History of Modern War*. New York: Oxford University Press, 74–93.

French, S. (1974). 'The Cemetery as Cultural Institution: The Establishment of Mount Auburn and the "Rural Cemetery" Movement', *American Quarterly*, 26/1: 37–59.

Fussell, P. (2013). *The Great War and Modern Memory*. Oxford and New York: Oxford University Press.

Gabriel, R. A. (2013). *Between Flesh and Steel: A History of Military Medicine from the Middle Ages to the War in Afghanistan*. Washington: Potomac Books.

Galinié, M. (1996). 'Le Passage de la nécropole au cimetière: Les Habitants des villes et leurs morts, du début de la christianisation à l'an mil', *Archéologie du cimetière chrétien*, 11: 17–22.

Galmozzi, L. (1986). *Monumenti alla libertà: Antifascimo, Resistenza e pace nei monumenti italiani dal 1945 al 1985*. Milan: La Pietra.

Galula, D. (2006). *Counterinsurgency Warfare: Theory and Practice*. New York: Praeger.

Gammage, B. (1974). *The Broken Years: Australian Soldiers in the Great War*. Canberra: Australian National University Press.

Garin, E. (1975). *Italian Humanism*. Westport: Greenwood Press.

Garton, S. (2006). 'Longing for War: Nostalgia and Australian Returned Soldiers After the First World War', in T. G Ashplant, G. Dawson, and M. Roper (eds), *Commemorating War: The Politics of Memory* New Brunswick and London: Transactions Publishers, 222–39.

Gasbarra, F. (2017). 'PDD-25 and the Genocide in Rwanda: Why Not a Task for the United States?', Insight and Analysis, Wilson Center, 18 September. https://www.wilsoncenter.org/blog-post/pdd-25-and-the-genocide-rwanda-why-not-task-for-the-united-states.

Gat, A. (2011). 'The Changing Character of War', in H. Strachan and S. Scheipers (eds), *The Changing Character of War*. Oxford: Oxford University Press, 27–47.

Gelpi, C. (2006). 'The Costs of War: How Many Casualties Will Americans Tolerate?: Misdiagnosis', *Foreign Affairs*, 85/1: 139–42.

Gelpi C., P. Feaver, and J. Reifler (2009). *Paying the Human Costs of War: American Public Opinion and Casualties in Military Conflicts*. Princeton: Princeton University Press.

Gentili, A. (2008). *Il diritto di guerra. DE IURE BELLI LIBRI III, 1598*. Milan: Giuffrè.

Gentry, J. A. (1998). 'Military Force in an Age of National Cowardice', *Washington Quarterly*, 21/4: 179–91.

Gherardi, R. (2014). 'Per una scienza globale della guerra', *Il pensiero politico*, 47/3: 295–318.

Giddens, A. (1991). *Modernity and Self-Identity: Self and Identity in the Late Modern Age*. Stanford: Stanford University Press.

Gill, C. J., and G. C. Gill (2005). 'Nightingale in Scutari: Her Legacy Reexamined', *Clinical Infectious Diseases*, 40/12: 1799–1805.

Gillis, J. R. (1994). 'Memory and Identity: The History of a Relationship', in J. R. Gillis (ed.), *Commemorations: The Politics of National Identity*. Princeton: Princeton University Press, 3–24.

Giustozzi, A. (2007). *Koran, Kalashnikov and Laptop: The Neo-Taliban Insurgency in Afghanistan*. London: Hurst.

Glazer, N. (1996). 'Monuments in an Age without Heroes', *Public Interest*, 123 (Spring), 22–39.

Goebel, S. (2004). 'The "Sleeping Dead" in Interwar Germany and Britain', *Journal of Contemporary History*, 39/4: 487–501.

Goebel, S. (2006). *The Great War and Medieval Memory: War, Remembrance and Medievalism in Britain and Germany, 1914–49*. Cambridge: Cambridge University Press.

Goldberg, M. S. (2014). 'Updated Death and Injury Rates of US Military Personnel during the Conflicts in Iraq and Afghanistan', Working Paper 2014–08, Congressional Budget Office, December, Washington, https://www.cbo.gov/sites/default/files/113th-congress-2013-2014/workingpaper/49837-Casualties_WorkingPaper-2014-08_1.pdf.

Goldfield, D. (2011). *America Aflame: How the Civil War Created a Nation*. New York: Bloomsbury Press.

Goozen, S. van. (2021). *Distributing the Harm of Just Wars: In Defence of an Egalitarian Baseline*. New York: Routledge.

Gorer, G. (1965). *Death, Grief, and Mourning in Contemporary Britain*. Garden City, NY: Doubleday & Co.

Graham, E.-J. (2006). *The Burial of the Urban Poor in Italy in the Late Roman Republic and Early Empire*. Oxford: Archaeopress.

Grant, S. M. (2005). 'Raising the Dead: War, Memory and American National Identity', *Nations and Nationalism*, 11/4: 509–29.

Gray, C. H. (1997). *Postmodern War: The New Politics of Conflict*. London: Routledge.

Gray, C. S. (2005). *Another Bloody Century: Future Warfare*. London: Phoenix.

Gray, C. S. (2006). *Out of the Wilderness: Prime-Time for Strategic Culture*. Defense Threat Reduction Agency, Advanced Systems and Concepts Office, 31 October. http://www.fas.org/irp/agency/dod/dtra/stratcult-out.pdf.

Gray, H. C. (2003). 'Posthuman Soldiers in Postmodern War', *Body & Society*, 9/4: 215–26.

Greenberg, A. (1989). 'Lutyens's Cenotaph', *Journal of the Society of Architectural Historians*, 48/1: 5–23.

Greenblatt, S. (1980). *Renaissance Self-Fashioning: From More to Shakespeare*. Chicago: University of Chicago Press.

Greenwood, J. T., and F. C. Berry Jr (2005). *Medics at War: Military Medicine from Colonial Times to the 21st Century*. Annapolis, MD: Naval Institute Press.

Gregory, T. (2015). 'Dismembering the Dead: Violence, Vulnerability and the Body in War', *European Journal of International Relations*, 22/4: 944–65.

Gregory, T. (2019). 'The Costs of War: Condolence Payments and the Politics of Killing Civilians', *Review of International Studies*, 46/1: 156–76.

Grider, N. (2007). '"Faces of the Fallen" and the Dematerialization of US War Memorials', *Visual Communication*, 6/3: 265–79.

Grotelueschen, M. E. (2007). *The AEF Way of War: The American Army and Combat in World War I*. New York: Cambridge University Press.

Gudmundsson, B. I. (1993). *On Artillery*. Westport: Praeger.

Guldi, J., and D. Armitage (2014). *The History Manifesto*. Cambridge: Cambridge University Press.

Gusfield, J., and J. Michalowicz. (1984). 'Secular Symbolism: Studies of Ritual, Ceremony, and the Symbolic Order in Modern Life', *Annual Review of Sociology*, 10: 417–35.

Hagerman, E. (1988.) *The American Civil War and the Origins of Modern Warfare: Ideas, Organization, and Field Command.* Bloomington and Indianapolis: Indiana University Press.

Hagopian, P. (2009). *The Vietnam War in American Memory: Veterans, Memorials, and the Politics of Healing.* Amherst: University of Massachusetts Press.

Halbwachs, M. (1980). *The Collective Memory.* New York: Harper & Row.

Halbwachs, M. (1992). *On Collective Memory*, ed., trans., and with an introduction by L. C. Coser. Chicago: Chicago University Press.

Hale, J. R. (1985). *War and Society in Renaissance Europe, 1450–1620.* Baltimore: Johns Hopkins University Press.

Hall, R. (2016). *Atlantic Politics, Military Strategy and the French and Indian War.* Basingstoke: Palgrave Macmillan.

Hanson, V. D. (1989). *The Western Way of War: Infantry Battle in Classical Greece.* New York: Alfred A. Knopf.

Hanson, V. D. (2001a). *Carnage and Culture: Landmark Battles in the Rise to Western Power.* New York: Random House.

Hanson, V. D. (2001b). *Why the West Has Won: Carnage and Culture from Salamis to Vietnam.* London: Faber and Faber.

Harari, Y. N. (2008). *The Ultimate Experience: Battlefield Revelation and the Making of Modern War Culture, 1450–2000.* London: Palgrave.

Harrison, J. P. (1993). 'History's Heaviest Bombing', in J. S. Werner and L. D. Huynh (eds), *The Vietnam War: Vietnamese and American Perspectives.* New York: M. E. Sharpe Inc., 130–9.

Harrison, R. P. (2003). *The Dominion of the Dead.* Chicago: University of Chicago Press.

Hart, P. (2008). *The Somme: The Darkest Hour on the Western Front.* New York: Pegasus Books.

Hass, K. A. (1998). *Carried to the Wall: American Memory and the Vietnam Veterans War.* Berkeley and Los Angeles: University of California Press.

Hastings, M. (2007). *Retribution: The Battle for Japan, 1944–45.* New York: Alfred A. Knopf.

Hawley, T. M. (2002). 'Bodies and Border Practices: The Search for American MIAs in Vietnam', *Body & Society*, 8/3: 49–69.

Hayashi, N. (2020). *Military Necessity: The Art, Morality and Law of War.* Cambridge: Cambridge University Press.

Herrera, R. A. (2001). 'Self-Governance and the American Citizen as Soldier, 1775-1861', *Journal of Military History*, 65/1: 21–52.

Hess, E. J. (1997). 'Tactics, Trenches, and Men in the Civil War', in S. Förster and J. Nagler (eds), *On the Road to Total War: The American Civil War and the German Wars of Unification, 1861–1871.* New York: Cambridge University Press, 481–96.

Heuser, B. (2007) 'Introduction', in C. von Clausewitz, *On War.* Oxford and New York: Oxford University Press, pp.vii -xxxii.

Heuser, B. (2010) 'Small Wars in the Age of Clausewitz: The Watershed between Partisan War and People's War', *Journal of Strategic Studies*, 33/1: 139–62.

Heyningen, E. van, and P. Merrett (2002). '"The Healing Touch": The Guild of Loyal Women of South Africa 1900–1912', *South African Historical Journal*, 47/1: 24–50.

Hijiya, J. A. (1983). 'American Gravestones and Attitudes toward Death: A Brief History', *Proceedings of the American Philosophical Society*, 127/5: 339–63.

Hobbes, T. (2012). *Leviathan*, ed. N. Malcom. Oxford: Oxford University Press, vol. II.

Hobsbawm, E. (1983). 'Mass-Producing Traditions: Europe 1870–1914', in E. Hobsbawm and T. Ranger (eds), *The Invention of Tradition*. Cambridge: Cambridge University Press, 263–307.

Hoffman, F. G. (2007). *Conflict in the 21st Century: The Rise of Hybrid Wars*. Arlington: Potomac Institute for Policy Studies.

Holden Reid, B. (2006). *The Civil War and the Wars of the Industrial Revolutions*. New York: HarperCollins.

Holmes, R. (1989). *Acts of War: The Behavior of Men in Battle*. New York: Free Press.

Holmes, R. (2005). 'Battle: The Experience of Modern Combat', in C. Townshend, (ed.) *The Oxford History of Modern War*. New York: Oxford University Press, 224–44.

Holmes, R. (2011). *Soldiers: Army Lives and Loyalties from Redcoats to Dusty Warriors*. London: Harper Press.

Hope, V. M. (2018). 'Dulce et decorum est pro patria mori': The Practical and Symbolic Treatment of the Roman War Dead', *Mortality*, 23/1: 35–49.

Horne, A. (1981). *The Price of Glory: Verdun 1916*. London: Penguin.

Horowitz, M. C. (2017). 'Military Robotics, Autonomous Systems, and the Future of Military Effectiveness', in D. Reiter (ed.), *The Sword's Other Edge: Trade-offs in the Pursuit of Military Effectiveness*. Cambridge: Cambridge University Press, 161–96.

Howard, M. (1961). *The Franco-Prussian War: The German Invasion of France, 1870–1871*. New York: Macmillan.

Howard, M. (1976). *War in European History*. Oxford: Oxford University Press.

Howard, M. (1979). 'The Forgotten Dimension of Strategy', *Foreign Affairs*, 57/5: 975–86.

Howard, M. (1986). 'Men against Fire: The Doctrine of the Offensive in 1914', in P. Paret (ed.), with the collaboration of G. A. Craig and F. Gilbert, *Makers of Modern Strategy: From Machiavelli to the Nuclear Age*. Princeton: Princeton University Press, 510–26.

Howard, M. (2000). *The Invention of Peace: Reflections on War and International Order*. New Haven and London: Yale University Press.

Huizinga, J. (1922). *The Waning of the Middle Ages: A Study of the Forms of Life, Thought and Art in France and the Netherlands in the Fourteenth and Fifteenth Centuries*. Harmondsworth: Penguin.

Hull, I. V. (2005). *Absolute Destruction: Military Culture and the Practices of War in Imperial Germany*. Ithaca, NY: Cornell University Press.

Huntington, S. P. (1957). *The Soldier and the State: The Theory and Politics of Civil-Military Relations*. Cambridge, MA: Belknap Press.

Hutchinson, J. (2017). *Nationalism and War*. Oxford: Oxford University Press.

Hynes, S. (1990). *A War Imagined. The Great War and English Culture*. London: Bodley Head.

Ignatieff, M. (1984). 'Soviet War Memorials', *History Workshop Journal*, 17/1: 157–63.

Illich, I. (1976). *Medical Nemesis: The Expropriation of Health*. London: Marion Boyars.

Inglis, K. S. (2008). *Sacred Places: War Memorials in the Australian Landscape*. Melbourne: Melbourne University Press.

Instructions for the Government of Armies of the United States in the Field (*Lieber Code*), 24 April 1863. https://ihl-databases.icrc.org/ihl/INTRO/110.

ISAF (2009). 'ISAF Commander's Counterinsurgency Guidance'. Kabul, Afghanistan, 26 August; http://www.nato.int/isaf/docu/official_texts/counterinsurgency_guidance.pdf.

Isnenghi, M. (2005). *Le guerre degli italiani: Parole, immagini, ricordi 1848–1945*. Bologna: Mulino.

Jalland, P. (2010). *Death in War and Peace: A History of Loss and Grief in England, 1914–1970*. Oxford: Oxford University Press.

Janney, C. E. (2008). *Burying the Dead but Not the Past: Ladies' Memorial Associations & the Lost Cause*. Chapel Hill: University of North Carolina Press.
Jesi, F. (1979). *Cultura di destra*. Milan: Garzanti.
Joshel, S. R. (2010). *Slavery in the Roman World*. Cambridge: Cambridge University Press.
Juergensmeyer, M. (2020). *God at War: Meditation on Religion and Warfare*. New York: Oxford University Press.
Kaempf, S. (2018). *Saving Soldiers or Civilians? Casualty-Aversion versus Civilian Protection in Asymmetric Conflicts*. Cambridge: Cambridge University Press.
Kahn, P. W. (2002). 'The Paradox of Riskless Warfare', *Philosophy and Public Policy Quarterly*, 22/3: 2–8.
Kaldor, Mary (2012). *New and Old Wars: Violence in a Global Era*. Stanford: Stanford University Press.
Kalyvas, S. N. (1999). 'Wanton and Senseless? The Logic of Massacres in Algeria', *Rationality and Society*, 11/3: 243–85.
Kamm, F. M. (2020). *Almost Over: Aging, Dying, Dead*. New York: Oxford University Press.
Kant, I. (2012). *Groundwork of the Metaphysics of Morals*, ed. and trans. Mary Gregor and Jens Timmerman. Cambridge: Cambridge University Press.
Kantorowicz, E. H. (1997). *The King's Two Bodies: A Study in Medieval Political Theology*. Princeton: Princeton University Press.
Kaplan, R. D. (2005). *Balkan Ghosts: A Journey through History*. New York: Picador.
Kattago, S. (2009). 'War Memorials and the Politics of Memory: The Soviet War Memorial in Talllin', *Constellations*, 16/1: 150–66.
Katz, Y. (2014). *The Tombstone in Israel's Military Cemetery since 1948: Israel's Transition from Collectivism to Individualism*. Berlin and Boston: de Gruyter.
Katzenstein, P. J. (1996) (ed.). *The Culture of National Security: Norms and Identity in World Politics*. New York: Columbia University Press.
Kearl, M. C. (1989). *Endings: A Sociology of Death and Dying*. New York: Oxford University Press.
Keegan, J. (1993). *A History of Warfare*. London: Hutchinson.
Keegan, J. (1999a). *War and our World: The Reith Lectures 1998*. London: Pimlico.
Keegan, J. (1999b). *The First World War*. New York: Alfred A. Knopf.
Keegan, J. (2004). *The Face of Battle: A Study of Agincourt, Waterloo and the Somme*. London: Pimlico.
Keegan, J., and R. Holmes (1985). *Soldiers: A History of Men in Battle*. London: Hamish Hamilton.
Keene, J. (2010). 'Bodily Matters Above and Below Ground: The Treatment of American Remains from the Korean War', *Public Historian*, 32/1: 59–78.
Kellehear, A. (2007). *A Social History of Dying*. Cambridge: Cambridge University Press.
Kemp, B. J. (2006). *Ancient Egypt: Anatomy of a Civilization*. London: Routledge.
Kestnbaum, M. (2005). 'Mars Revealed: The Entry of Ordinary People into War among States', in J. Adams, E. S. Clemens, and A. S. Orloff (eds), *Remaking Modernity: Politics, History, and Sociology*. Durham NC: Duke University Press, 249–85.
Kier, E. (1997). *Imagining War: French and British Military Doctrine between the Wars*. Princeton: Princeton University Press.
Kilcullen, D. (2006). '"Twenty-Eight Articles": Fundamentals of Company-Level Counterinsurgency', *Military Review*, 83/3: 103–8.
Kilcullen, D. (2009). *The Accidental Guerrilla: Fighting Small Wars in the Midst of a Big One*. New York: Oxford University Press.

Kimble, J. C. (2016). 'Spectral Soldiers: Domestic Propaganda, Visual Cultural, and Images of Death on the World War II Home Front', *Rhetoric & Public Affairs*, 19/4: 535–70.
King, A. (1998). *Memorials of the Great War in Britain: The Symbolism and Politics of Remembrance*. London and New York: Bloomsbury.
King, A. (2010). 'The Afghan War and "Postmodern" Memory: Commemoration and the Dead of Helmand', *British Journal of Sociology*, 61/1: 1–25.
Kinsella, H. M. (2011.) *The Image before the Weapon: A Critical History of the Distinction between Combatant and Civilian*. Ithaca, NY: Cornell University Press.
Kissinger, H. A. (1969). 'The Viet Nam Negotiations', *Foreign Affairs*, 47/2: 211–34.
Kissinger, H. (1994). *Diplomacy*. New York: Simon & Schuster.
Kohn, R. H. (1981). 'The Social History of the American Soldier: A Review and Prospectus for Research', *American Historical Review*, 86/3: 553–67.
Koselleck, R. (2002). *The Practice of Conceptual History: Timing History, Spacing Concepts*. Stanford: Stanford University Press.
Kozak, W. (2009). *LeMay: The Life and Wars of General Curtis LeMay*. Washington: Regnery.
Kramer, A. (2007). *Dynamic of Destruction: Culture and Mass Killing in the First World War*. Oxford: Oxford University Press.
Krepinevich, A. F., Jr (1988). *The Army and Vietnam*. Baltimore and London: Johns Hopkins University Press.
Krieg, A., and J.-M. Rickli (2019). *Surrogate Warfare: The Transformation of War in the Twenty-First Century*. Washington: Georgetown University Press.
Kriner, D. L., and F. X. Shen (2010). *The Casualty Gap: The Causes and Consequences of American Wartime Inequalities*. Oxford and New York: Oxford University Press.
Lacina, B. (2006). 'Explaining the Severity of Civil Wars', *Journal of Conflict Resolution*, 50/2: 276–89.
Lacina, B., and N. P. Gleditsch. (2005). 'Monitoring Trends in Global Combat: A New Dataset of Battle Deaths', *European Journal of Population*, 21: 145–66.
Lacina, B., N. P. Gleditsch, and B. Russett (2006). 'The Declining Risk of Death in Battle', *International Studies Quarterly*, 50/3: 673–80.
Laderman, G. (1996). *The Sacred Remains: American Attitudes toward Death, 1799–1883*. New Haven: Yale University Press.
Laderman, G. (1997). 'The Body Politic and the Politics of Two Bodies: Abraham Lincoln and Mary Todd Lincoln in Death', *Prospects*, 22 (October), 109–32.
Lambert, N. A. (2012). *Planning Armageddon: British Economic Warfare and the First World War*. Cambridge, MA: Harvard University Press.
Laqueur, T. W. (1983). 'Bodies, Death and Pauper Funerals', *Representations*, 1 (February), 109–31.
Laqueur, T. W. (1994). 'Memory and Naming in the Great War', in J. R. Gillis (ed.), *Commemorations: The Politics of National Identity*. Princeton: Princeton University Press, 150–67.
Laqueur, T. W. (1996). 'Names, Bodies, and the Anxiety of Erasure', in T. R. Schatzki and W. Natter (eds), *The Social and Political Body*. New York: Guilford Press, 123–41.
Laqueur, T. W. (2001). 'Spaces of the Dead', *Ideas: from the National Humanities Center*, 8/2: 3–16.
Laqueur, T. W. (2015). *The Work of the Dead*. Princeton: Princeton University Press.
Larson, E. V., and B. Savych (2005). *American Support for US Operations from Mogadishu to Baghdad*. Santa Monica: RAND.

Lauwers, M. (2005). *Naissance du cimetière: Lieux sacrés et terre des morts dans l'Occident médiéval*. Paris: Aubier.
Le Goff, J. (1984). *The Birth of the Purgatory*. London: University of Chicago Press.
Lebow, R. N. (2008). *A Cultural Theory of International Relations*. Cambridge: Cambridge University Press.
Lee, W. E. (2011). 'Warfare and Culture', in W. E. Lee (ed.), *Warfare and Culture in World History*. New York: New York University Press, 1–11.
Legacey, E.-M. (2019). *Making Space for the Dead: Catacombs, Cemeteries, and the Reimagining of Paris, 1780–1830*. Ithaca, NY: Cornell University Press.
Legro, J. W. (1995). *Cooperation under Fire: Anglo-German Restraint during World War II*. Ithaca, NY: Cornell University Press.
Levi, P. (1959). *If this is a Man*. New York: Orion Press.
Levy, Y. (2012). *Israel's Death Hierarchy: Casualty Aversion in a Militarized Democracy*. New York: New York University Press.
Lewis, A. R. (2007). *The American Culture of War: A History of US Military Force from World War II to Operation Iraqi Freedom*. New York: Routledge.
Lewis, A. R. (2012). *The American Culture of War: A History of US Military Force from World War II to Operation Enduring Freedom*. New York: Routledge.
Lincoln, A. (2013). 'The Gettysburg Address', in T. Ball (ed.), *Lincoln: Political Writings and Speeches*. Cambridge: Cambridge University Press, 191–2.
Lincoln, B. (1991). *Death, War, and Sacrifice: Studies in Ideology and Practice*. Chicago: University of Chicago Press.
Linenthal, E. T. (1991). *Sacred Ground: Americans and their Battlefields*. Urbana: University of Illinois Press.
Llewellyn, N. (1991). *The Art of Death: Visual Culture in the English Death Ritual* c.*1500*–c.*1800*. London: Reaktion Books.
Lloyd, D. W. (1998). *Battlefield Tourism: Pilgrimage and Commemoration of the Great War in Britain, Australia and Canada, 1919–1939*. Oxford and New York: Berg.
Lomnitz, C. (2005). *Death and the Idea of Mexico*. New York: Zone Books.
Longworth, P. (2003). *The Unending Vigil: The History of the Commonwealth War Graves Commission*. Barnsley: Leo Cooper.
Loraux, N. (2006). *The Invention of Athens: The Funeral Oration in the Classical City*. New York: Zone Books.
Lowenthal, D. (1985). *The Past is a Foreign Country*. Cambridge: Cambridge University Press.
Luard, E. (1986). *War in International Society: A Study in International Sociology*. London: I. B. Tauris & Co. Ltd.
Luard, E. (1988). *The Blunted Sword: The Erosion of Military Power in Modern World Politics*. New York: New Amsterdam.
Lüdtke, A. (1997). 'Histories of Mourning: Flowers and Stones for the War Dead, Confusion for the Living—Vignettes from East and West Germany', in G. Sider and G. Smith (eds), *Between History and Histories: The Making of Silences and Commemorations*. Toronto: University of Toronto Press, 149–79.
Lujala, P. (2009). 'Deadly Combat over Natural Resources: Gems, Petroleum, Drugs, and the Severity of Armed Conflict', *Journal of Conflict Resolution*, 53/1: 50–71.
Luttwak, E. N. (1994). 'Where Are the Great Powers? At Home with the Kids', *Foreign Affairs*, 73/4: 23–8.
Luttwak, E. N. (1995). 'Toward Post-Heroic Warfare', *Foreign Affairs*, 74/3: 109–22.

Lyall, J., and I. Wilson III (2009). 'Rage against the Machines: Explaining Outcomes in Counterinsurgency Wars', *International Organization*, 63/1: 67–106.
Lynn, J. A. (2003). *Battle: A History of Combat and Culture*. Boulder, CO: Lynne Rienner.
Lynn, J. A. (2005). 'States in Conflict', in G. Parker (ed.), *The Cambridge History of Warfare*. New York: Cambridge University Press, 167–88.
Lynn, J. A. (2011). 'The Battle of Forbearance, 1660–1789', in W. E. Lee (ed.), *Warfare and Culture in World History*. New York: New York University Press, 89–113.
McCartney, H. (2010) 'The Military Covenant and the Civil–Military Contract in Britain', *International Affairs*, 86/2: 411–28.
McCloskey, D. N. (2016). *Bourgeois Equality: How Ideas, not Capital or Institutions, Enriched the World*. Chicago: University of Chicago Press.
MacEachin, D. J. (1998). *The Final Months of the War with Japan. Signals Intelligence, US Invasion Planning, and the A-Bomb Decision*. Langley, VA: Center for the Study of Intelligence, https://oregondigital.org/sets/easia/oregondigital:df72db551#page/34/mode/1up.
Macfarlane, A. (1978.) *The Origins of English Individualism*. Oxford: Blackwell.
Machiavelli, N. (1998). *The Prince*, trans. H. C. Mansfield. Chicago and London: University of Chicago Press.
Machiavelli, N. (2001). *The Art of War*. Cambridge, MA: Da Capo Press.
Mack, A. (1975). 'Why Big Nations Lose Small Wars: The Politics of Asymmetric Conflict', *World Politics*, 27/2: 175–200.
McMahan, J. (2002). *The Ethics of Killing: Problems at the Margins of Life*. New York: Oxford University Press.
McMahan, J. (2009). *Killing in War*. New York: Oxford University Press.
McMains, H. F. (2000). *The Death of Oliver Cromwell*. Lexington: University Press of Kentucky.
McNeill, W. H. (1982). *The Pursuit of Power: Technology, Armed Force, and Society since* AD *1000*. Chicago: University of Chicago Press.
McNeill, W. H. (1995). *Keeping Together in Time: Dance and Drill in Human History*. Cambridge, MA: Harvard University Press.
McPhedran, I. (2013). 'The First 75 of Lost Australian Soldiers at Fromelles Positively Identified', *Daily Telegraph*, 16 March.
McSorley, K. (2014). 'Towards an Embodied Sociology of War', *Sociological Review*, 62/2: 107–28.
McWhiney, G., and P. D. Jamieson (1982). *Attack and Die: Civil War Military Tactics and the Southern Heritage*. Tuscaloosa: University of Alabama Press.
Mailer, N. (1967) *Why Are We in Vietnam?* New York: G. P. Putnam's Sons.
Malešević, S. (2010a). *The Sociology of War and Violence*. New York: Cambridge University Press.
Malešević, S. (2010b). 'How Pacifist Were the Founding Fathers?: War and Violence in Classical Sociology', *European Journal of Social Theory*, 13/2: 193–212.
Mallett, D. R. (2013). 'Monuments and Memorials', in S. C. Tucker (ed.), *The Encyclopedia of the American–Mexican War, Volume I: A-L*. Santa Barbara: ABC-CLIO, 443–4.
Malone, H. (2017). *Architecture, Death and Nationhood: Monumental Cemeteries of Nineteenth-Century Italy*. London: Routledge.
Malone, H. (2019a). 'New Life in the Modern Cultural History of Death', *Historical Journal*, 62/3: 833–52.
Malone, H. (2019b). 'The Republican Legacy of Italy's Fascist Ossuaries of the First World War', *Modern Italy*, 24/2: 199–217.

Mann, M. (2012a). *The Sources of Social Power, Vol. 1: A History of Power from the Beginning to* AD *1760*. Cambridge: Cambridge University Press.
Mann, M. (2012b). *The Sources of Social Power, Vol. 2: The Rise of Classes and Nation-States, 1760–1914*. Cambridge: Cambridge University Press.
Markovits, S. (2012). 'On the Crimean War and the Charge of the Light Brigade', in D. F. Felluga (ed.), *BRANCH: Britain, Representation and Nineteenth-Century History* (April), https://www.branchcollective.org/?ps_articles=stefanie-markovits-on-the-crimean-war.
Marshall, S. L. A. (1947). *Men against Fire: The Problem of Battle Command*. New York: William Morrow and Company.
Matloff, M. (1994). *Strategic Planning for Coalition Warfare, 1943–1944*. Washington: Center for Military History, United States Army.
Maude, F. N. (1905). *Notes on the Evolution of Infantry Tactics*. London: William Clowes & Sons.
May, W. F. (1972). 'The Sacral Power of Death in Contemporary Experience', *Social Research*, 39/3: 463–88.
Mayo, J. M. (1988). 'War Memorials as Political Memory', *Geographical Review*, 78/1: 62–75.
Mazzantini, C. (1995). *A cercar la bella morte*. Venice: Marsilio.
Mbembe, A. (2003). 'Necropolitics', *Public Culture*, 15/1: 11–40.
Meigs, M. (1994). 'La Mort et ses enjeux: L'Utilisation des corps des soldats américains lors de la Première Guerre mondiale', *Guerres mondiales et conflits contemporains*, 175 (July), 135–46.
Merridale, C. (1999). 'War, Death, and Remembrance in Soviet Russia', in J. Winter, and E. Sivan (eds), *War and Remembrance in the Twentieth Century*. Cambridge and New York: Cambridge University Press, 61–83.
Metcalf, P., and R. Huntington (1991). *Celebrations of Death: The Anthropology of Mortuary Ritual*. Cambridge: Cambridge University Press.
Middlebrook, M. (1984). *The First Day on the Somme*. London: Penguin.
Mitchell, R. (1997). *Civil War Soldiers*. New York: Penguin.
Mitzen, J. (2006). 'Ontological Security in World Politics: State Identity and the Security Dilemma', *European Journal of International Relations*, 12/3: 341–70.
Mock, S. J. (2012). *Symbols of Defeat in the Construction of National Identity*. Cambridge and New York: Cambridge University Press.
Mombauer, A. (2001). *Helmuth von Moltke and the Origins of the First World War*. Cambridge: Cambridge University Press.
Moran, D. (2007). 'The Instrument: Clausewitz on Aims and Objectives in War', in H. Strachan, and A. Herberg-Rothe (ed.), *Clausewitz in the Twenty-First Century*. Oxford: Oxford University Press, 91–106.
Morgan, O., T. B. Morris, and D. van Alphen (2006) (eds). *Management of Dead Bodies after Disasters: A Field Manual for First Responders*. Washington: Pan American Health Organization.
Morgan, S. (1998). 'Memory and the Merchants: Commemoration and Civic Identity', *International Journal of Heritage Studies*, 4/2: 103–13.
Morgenthau, H. J. (1965). *Scientific Man vs Power Politics*. Chicago: University of Chicago Press.
Moriarty, C. (1997). 'Private Grief and Public Remembrance: British First World War Memorials', in M. Evans and K. Lunn (eds), *War and Memory in the Twentieth Century*. Oxford: Berg, 125–42.

Morris, C. (1972). *The Discovery of the Individual, 1050–1200*. London and New York: Harper and Row.
Morris, I. (1992). *Death-Ritual and Social Structure in Classical Antiquity*. Cambridge: Cambridge University Press.
Morrissey, R. (2014). *The Economy of Glory: From Ancien Régime France to the Fall of Napoleon*. Chicago and London: University of Chicago Press.
Moskos, C. (2001). 'What Ails the All-Volunteer Force: An Institutional Perspective', *Parameters*, 31/1: 29–47.
Moskos, C. C., Jr (1970). *The American Enlisted Man: The Rank and File in Today's Military*. New York: Russell Sage Foundation.
Mosse, G. L. (1979). 'National Cemeteries and National Revival: The Cult of the Fallen Soldiers in Germany', *Journal of Contemporary History*, 14/1: 1–20.
Mosse, G. L. (1990). *Fallen Soldiers: Reshaping the Memory of the World Wars*. Oxford: Oxford University Press.
Moyn, S. (2021). *Humane: How the United States Abandoned Peace and Reinvented Peace*. New York: Farrar, Straus and Giroux.
Mueller, J. (1973). *War, Presidents, and Public Opinion*. New York: Wiley.
Mueller, J. (1989). *Retreat from Doomsday*. New York: Basic Books.
Mueller, K. P. (2000). 'Politics, Death, and Morality in US Foreign Policy', *Aerospace Power Journal* (Summer), 12–16, https://apps.dtic.mil/sti/pdfs/ADA517080.pdf.
Mueller J. (2011). 'The Iraq Syndrome Revisited: US Intervention, from Kosovo to Libya', foreignaffairs.com, March 28.
Münkler, H. (2005). *The New Wars*. Cambridge: Polity.
Murray, W. (1996). 'American Civil War', in R. Cowley and G. Parker (ed.), *The Osprey Companion to Military History* London: Houghton Mifflin Company, 15–19.
Murray, W. A. (2005). 'The Industrialization of War', in G. Parker (ed.), *The Cambridge History of Warfare*. New York: Cambridge University Press, 219–48.
Mytum, H. (1989). 'Public Health and Private Sentiment: The Development of Cemetery Architecture and Funerary Monuments from the Eighteen Century Onwards', *World Archaeology*, 21/2: 283–97.
National Review (1981). 'Stop that Monument', 33, 18 September, p. 1064.
Nef, J. U. (1950). *War and Human Progress: An Essay on the Rise of Industrial Civilization*. Cambridge, MA: Harvard University Press.
Neff, J. R. (2004). *Honoring the Civil War Dead: Commemoration and the Problem of Reconciliation*. Lawrence: University Press of Kansas.
Neiberg, M. S. (2006). *Soldiers' Lives through History: The Nineteenth Century*. Westport and London: Greenwood Press.
Neillands, R. (2001). *The Bomber War: Arthur Harris and the Allied Bomber Offensive 1939–1945*. London: John Murray.
Nightingale, F. (2011). 'Army Sanitary Administration and its Reform under the Late Lord Herbert', in L. McDonald (ed.), *Florence Nightingale on War and the War Office, Volume 15 of the Collected Works of Florence Nightingale*. Waterloo: Wilfrid Laurier University Press, 342–51.
Nitze, P. H. (1989). *From Hiroshima to Glasnost: At the Center of Decision—A Memoir*. New York: Grove Weidenfeld.
Nive, B. (2002). *Facing the Nazi Past: United Germany and the Legacy of the Third Reich*. London and New York: Routledge.
Nora, P. (1984) (ed.). *Les Lieux de mémoire, Volume 1: La République*. Paris: Gallimard.
Nora, P. (1986) (ed.). *Les Lieux de mémoire, Volume 2: La Nation*. Paris: Gallimard.

Nora, P. (1992) (ed.). *Les Lieux de mémoire, Volume 3: Les France*. Paris: Gallimard.
Noy, D. (2000). '"Half-Burnt on an Emergency Pyre": Roman Cremations which Went Wrong', *Greece & Rome*, 47/2: 186–96.
Nudelman, F. (2004). *John Brown's Body: Slavery, Violence & the Culture of War*. Chapel Hill and London: University of North Carolina Press.
O'Connell, R. L. (1989). *Of Arms and Men: A History of War, Weapons, and Aggression*. New York: Oxford University Press.
Olick, J. K. (1999). 'Genre Memories and Memory Genres: A Dialogical Analysis of May 8, 1945 Commemorations in the Federal Republic of Germany', *American Sociological Review*, 64/3: 381–402.
Olsen, N. (2014). *History in the Plural: An Introduction to the Work of Reinhart Koselleck*. New York and Oxford: Berghahn Books.
Osborne, E. W. (2004). *Britain's Economic Blockade of Germany, 1914–1919*. London and New York: Frank Cass.
Overy, R. J. (1987). *The Air War, 1939–1945*. New York: Stein and Day.
Panofsky, E. (1964). *Tomb Sculpture: Four Lectures on its Changing Aspects from Ancient Egypt to Bernini*. New York: H. N. Abrams.
Pape, R. A. (2005). *Dying to Win: The Strategic Logic of Suicide Terrorism*. New York: Random House.
Parashar, S. (2013). 'What Wars and "War Bodies" Know about International Relations', *Cambridge Review of International Affairs*, 26/4: 615–30.
Paret, P. (1986). 'Clausewitz', in P. Paret (ed.), with the collaboration of G. A. Craig and F. Gilbert, *Makers of Modern Strategy: From Machiavelli to the Nuclear Age*. Princeton: Princeton University Press, 186–213.
Parey, A. (1825). 'The Works of that Famous Chirurgeon Ambrose Parey', *Retrospective Review*, 11/1: 46–66.
Parker, G. (1984). *The Thirty Years' War*. London: Routledge.
Parker, G. (2004). *The Army of Flanders and the Spanish Road, 1567–1659: The Logistics of Spanish Victory and Defeat in the Low Countries' Wars*. Cambridge: Cambridge University Press.
Parker, G. (2005). 'Dynastic War', in G. Parker (ed.), *The Cambridge History of Warfare*. New York: Cambridge University Press, 148–66.
Parker, G. (2007). 'The Limits to Revolutions in Military Affairs: Maurice of Nassau, the Battle of Nieuwpoort (1600), and the Legacy', *Journal of Military Affairs*, 71/2: 331–72.
Parker Pearson, M. (2000). *The Archaeology of Death and Burial*. College Station: Texas A&M University Press.
Pearlman, M. D. (2008). *Truman and MacArthur: Policy, Politics, and the Hunger for Honor and Renown*. Bloomington: Indiana University Press.
Perry, W. J. (1996). 'DoD's Re-evaluation of the Force Protection Posture', Defense Issues, 11/80, 16 September.
Pettersson, T., S. Högbladh, and M. Öberg (2019). 'Organized Violence, 1989–2018 and Peace Agreements', *Journal of Peace Research*, 56/4: 589–603.
Pfaff, W. (2004). *The Bullet's Song: Romantic Violence and Utopia*. New York: Simon & Schuster.
Phillips, J. B. (2007). 'The Changing Presentation of Death in the Obituary, 1899–1999', *Omega: Journal of Death and Dying*, 55/4: 325–46.
Philpott, W. (2009). *Bloody Victory: The Sacrifice on the Somme*. London: Abacus.
Philpott, W. (2014). *Attrition: Fighting the First World War*. London: Little, Brown.
Piehler, G. K. (1995). *Remembering War the American Way*. Washington: Smithsonian Institution Press.

Pinker, S. (2012). *The Better Angels of our Nature: A History of Violence and Humanity.* London: Penguin.

Pinker, S. (2018). *Enlightenment Now: The Case for Reason, Science, Humanism, and Progress.* London: Penguin.

Plumwood, V. (2008). 'Tasteless: Towards a Food-Based Approach to Death', *Environmental Values*, 17/3: 323–30.

Poole, R. M. (2009). *On Hallowed Ground: The Story of Arlington National Cemetery.* New York: Walker Publishing Company.

Price, R. (1997). *The Chemical Weapons Taboo.* Ithaca, NY: Cornell University Press.

Prost, A. (1997). 'Monuments to the Dead', in P. Nora (ed.), *Realms of Memory: The Construction of the French Past, Volume 2: Traditions.* New York: Columbia University Press, 307–32.

Prost, A. (2002). *Republican Identities in War and Peace: Representations of France in the 19th and 20th Centuries.* Oxford and New York: Berg.

Prost, A. (2011). 'The Military Cemeteries of the Great War, 1914–1940', *Le Mouvement social*, 237/4: 135–51.

Pyszczynski, T. A., S. Solomon, and J. Greenber (2003). *In the Wake of 9/11: The Psychology of Terror.* Washington: American Psychological Association.

Ragon, M. (1981). *L'Espace de la mort.* Paris: Albin Michel.

Ralph, W. W. (2006). 'Improvised Destruction: Arnold, LeMay, and the Firebombing of Japan', *War in History*, 13/4: 495–522.

Rearick, C. W. (1986). 'Funerals, State', in P.H. Hutton (ed.), *Historical Dictionary of the Third French Republic, 1870–1940*, Volume 1. Westport: Greenwood Press, 404–6.

Reimers, E. (1999). 'Death and Identity: Graves and Funerals as Cultural Communication', *Mortality*, 4/2: 147–66.

Reiter, D., and A. C. Stam (2002). *Democracies at War.* Princeton: Princeton University Press.

Richardson, F. M. (1985). 'Wellington, Napoleon, and the Medical Services', *Journal of the Royal Army Medical Corps*, 131/1: 9–15.

Richardson, R. (1989). 'Why Was Death so Big in Victorian England?', in R. Houlbrooke (ed.), *Death, Ritual and Bereavement.* London and New York: Routledge, 105–17.

Ridgway, M. B. (1967). *The Korean War.* Garden City, NY: Doubleday & Company.

Rigeade, C. (2008). 'Approche archéo-anthropologique des inhumations militaires', *Socio-anthropologie*, 22: 93–105.

Ritschl, A. (2005). 'The Pity of Peace: Germany's Economy at War, 1914–1918 and Beyond', in S. Broadberry and M. Harrison (eds), *The Economics of World War I.* New York: Cambridge University Press, 41–76.

Ritter, G. (1969). *The Sword and the Scepter: The Problem of Militarism in Germany, Volume 1: The Prussian Tradition 1740–1890.* Coral Gables, FL: University of Miami Press.

Robben, A. C. G. M. (2014). 'Governing the Disappeared-Living and the Disappeared-Dead', in F. Steputtat (ed.), *Governing the Dead: Sovereignty and the Politics of Dead Bodies.* Manchester: Manchester University Press, 143–62.

Roberts, M. (1992). *Gustavus Adolphus: A History of Sweden 1611–1632, Volume 2.* London: Longman.

Robertson, G. (2002). *Crimes against Humanity: The Struggle for Global Justice.* New York: New Press.

Robertson, J. I. (1988). *Soldiers Blue and Gray.* New York: Warner Books.

Robin, R. T. (1992). *Enclaves of America: The Rhetoric of Political Architecture Abroad.* Princeton: Princeton University Press.

Rogers, C. J. (2007). *Soldiers' Lives through History: The Middle Ages*. Westport and London: Greenwood Press.
Ropp, T. (2000). *War in the Modern World*. Baltimore and London: Johns Hopkins University Press.
Rosanvallon, P. (2011). *La Société des égaux*. Paris: Éditions du Seuil.
Rosenheim, J. L. (2013). *Photography and the American Civil War*. New York: Metropolitan Museum of Art.
Rostker, B. (2013). *Providing for the Casualties of War: The American Experience through World War II*. Santa Monica: RAND Corporation.
Rothenberg, G. E. (1986). 'Moltke, Schlieffen, and the Doctrine of Strategic Envelopment', in P. Paret (ed.), with the collaboration of G. A. Craig and F. Gilbert, *Makers of Modern Strategy: From Machiavelli to the Nuclear Age*. Princeton: Princeton University Press, 296–325.
Roy, I. (1977). 'The Army and its Critics in Seventeenth Century England', in B. Bond and I. Roy (eds), *War and Society: A Yearbook of Military History, Volume* 2.. London: Croom Helm, 141–50.
Rugg, J. (2000). 'Defining the Place of Burial: What Makes a Cemetery a Cemetery?', *Mortality*, 5/3: 259–75.
Salinger, J. D. (2014). *The Catcher in the Rye*. New York: Little, Brown and Company.
Samet, E. D. (2005). 'Leaving No Warriors Behind: The Ancient Roots of a Modern Sensibility', *Armed Forces & Society*, 31/4: 623–49.
Saunders, N. J. (2001). 'Apprehending Memory: Material Culture and War, 1919–1939', in P. Liddle, J. Bourne, and I. Whitehead (eds), *The Great World War 1914–45, Volume 2: The Peoples' Experience*. London: HarperCollins, 476–88.
Savage, K. (1997). *Standing Soldiers, Kneeling Slaves: Race, War, and Monument in Nineteenth Century America*. Princeton: Princeton University Press.
Scarinzi, F. (2021). 'Force Structure and Counterinsurgency Outcome: The Case of the Cyprus Emergency (1955–1959)', *Defence Studies* 21/2: 204–25.
Scarry, E. (1985). *The Body in Pain: The Making and Unmaking of the World*. New York: Oxford University Press.
Schwarz, G. (2003). 'La morte e la patria: l'Italia e i difficili lutti della seconda guerra mondiale', *Quaderni Storici*, 38/113: 551–87.
Scheck, R. (2005). '"They Are Just Savages": German Massacres of Black Soldiers from the French Army in 1940', *Journal of Modern History*, 77/2: 325–44.
Scheidel, W. (2018). *The Great Leveler: Violence and the History of Inequality from the Stone Age to the Twenty-First Century*. Princeton and Oxford: Princeton University Press.
Scheipers, S. (2018). *On Small War: Carl von Clausewitz and People's War*. Oxford: Oxford University Press.
Schelling, T. C. (1966). *Arms and Influence*. New Haven and London: Yale University Press.
Scheper-Hughes, N., and L. Wacquant (2002). *Commodifying Bodies*. London and Thousand Oaks, CA: Sage.
Schmitt, C. (2004). *Theory of the Partisan: A Commentary/Remark on the Concept of the Political*. East Lansing: Michigan State University Press.
Schwartz, B., and T. Bayma (1999). 'Commemoration and the Politics of Recognition: The Korean War Veterans Memorial', *American Behavioral Scientist*, 42/6: 946–67.
Searle, J. R. (1995). *The Construction of Social Reality*. New York: Free Press.
Semmel, S. (2000). 'Reading the Tangible Past: British Tourism, Collecting, and Memory after Waterloo', *Representations*, 69: 9–37.

Shah, N. (2017a). 'Death in the Details: Finding Dead Bodies at the Canadian War Museum', *Organization*, 24/4: 549–69.

Shah, N. (2017b). 'Gunning for War: Infantry Rifles and the Calibration of Lethal Force', *Critical Studies on Security*, 5/1: 81–104.

Shaw, M. (2003). *War and Genocide*. Cambridge: Polity.

Shaw, M. (2005). *The New Western Way of War*. Cambridge: Polity.

Sheffield, G. (2016). *Douglas Haig: From the Somme to Victory*. London: Aurum Press.

Sherman, D. (2014). *In a Strange Room: Modernism's Corpses and Mortal Obligation*. Oxford and New York: Oxford University Press.

Sherman, D. J. (1998). 'Bodies and Names: The Emergence of Commemoration in Interwar France', *American Historical Review*, 103/2: 443–66.

Sherry, V. (2018). 'Bare Death: The Falling Sacrifice of the First World War', in A. Houen and J.-M. Schram (eds), *Sacrifice and Modern War Literature: The Battle of Waterloo the War on Terror Houen*. Oxford: Oxford University Press, 92–102.

Shils, E. A., and M. Janowitz (1948). 'Cohesion and Disintegration in the Wehrmacht in World War II', *Public Opinion Quarterly*, 12/2: 280–315.

Showalter, D. E. (1975). *Railroads and Rifles: Soldiers, Technology and the Unification of Germany*. Hamden: Archon Books.

Showalter, D., and W. J. Astore (2007). *Soldiers' Lives through History: The Early Modern Period*. Westport and London: Greenwood Press.

Shy, J. W. (1976). *A People Numerous and Armed: Reflections on the Military Struggle for American Independence*. New York: Oxford University Press.

Siedentop, L. (2015). *Inventing the Individual: The Origins of Western Liberalism*. London: Penguin.

Silver, L. (2002). 'Shinning Armor: Emperor Maximilian, Chivalry, and War', in P. Cuneo (ed.), *Artful Armies, Beautiful Battles: Art and Warfare in Early Modern Europe*. Leiden: Brill, 61–86.

Singer, P. W. (2009). *Wired for War: The Robotics Revolution and Conflict in the Twenty-First Century*. New York: Penguin.

Sledge, M. (2005). *Soldier Dead*. New York: Columbia University Press.

Slim, H. (2007). *Killing Civilians: Method, Madness and Morality in War*. London: Hurst & Company.

Sloane, D. C. (1991). *The Last Great Necessity: Cemeteries in American History*. Baltimore and London: Johns Hopkins University Press.

Smith, H. (2005). 'What Costs Will Democracies Bear? A Review of Popular Theories of Casualty Aversion', *Armed Forces and Society*, 31/4: 487–512.

Smith, L., S. Audoin-Rouzeau, and A. Becker (2003). *France and the Great War, 1914–1918*. Cambridge: Cambridge University Press.

Smith, R. (2006). *The Utility of Force: The Art of War in the Modern World*. London: Penguin.

Smith, T. B. (2008). *The Golden Age of Battlefield Preservation: The Decade of the 1890s and the Establishment of America's First Five Military Parks*. Knoxville: University of Tennessee Press.

Smith, T. W. (2008). 'Protecting Civilians … or Soldiers? Humanitarian Law and the Economy of Risk in Iraq', *International Studies Perspectives*, 9/2: 144–64.

Snape, M. (2005). *The Redcoat and Religion: The Forgotten History of the British Soldier from the Age of Marlborough to the Eve of the First World War*. New York: Routledge.

Sokolowska-Paryż, M. (2012). *Reimagining the War Memorial, Reinterpreting the Great War: The Formats of British Commemorative Fiction*. Cambridge: Cambridge Scholars Publishing.

Sondhaus L. (2006). *Strategic Culture and Ways of War*. London: Routledge.
Sontag, S. (2003). *Regarding the Pain of Others*. New York: Farrar, Straus and Giroux.
Sophocles (1989). *Antigone*, trans. R. E. Braun. Oxford and New York: Oxford University Press.
Sozzi, M. (2009). *Reinventare la morte: Introduzione alla tanatologia*. Rome and Bari: Laterza.
Spars. S. (2014). 'Exposing Those Who Bury the Dead: A New Perspective in Modern Conflict Archaeology', in P. Cornish and N.J. Saunders (eds) , *Bodies in Conflict: Corporeality, Materiality and Transformation*. London and New York: Routledge, 103–21.
Springer, P. J. (2018) *Outsourcing War to Machines: The Military Robotics Revolution*. Santa Barbara and Denver: Praeger.
Squire, V. (2017). 'Governing Migration through Death in Europe and the US', *European Journal of International Relations*, 23/3: 513–32.
Stamp, G. (2007). *The Memorial to the Missing of the Somme*. London: Profile Books.
Stannard, D. E. (1977). *The Puritan Way of Death: A Study on Religion, Culture and Social Change*. New York: Oxford University Press.
Steere, E. (1948). 'Genesis of American Graves Registration 1861–1870', *Military Affairs*, 12/3: 149–61.
Stephenson, M. (2012). *The Last Full Measure: How Soldiers Die in Battle*. New York: Crown Publishers.
Stone, L. (1977). *The Family, Sex, and Marriage in England, 1500–1800*. New York: Harper & Row.
Stone, L. (1981). *The Past and the Present*. Boston, London, and Henley: Routledge & Kegan Paul.
Stouffer, S. A., A. A. Lumsdaine, M. H. Lumsdaine, R. M. Williams Jr, M. B. Smith, I. L. Janis, S. A. Star, and L. S. Cottrell Jr (1949). *The American Soldier: Combat and its Aftermath, Volume 2*. Princeton: Princeton University Press.
Stourzh, G. (1996). 'The Modern State: Equal Rights. Equalizing the Individual's Status and the Breakthrough of the Modern Liberal State', in J. Coleman (ed.), *The Individual in Political Theory and Practice*. Oxford: Oxford University Press, 303–27.
Strachan, H. (1983). *European Armies and the Conduct of War*. London: Routledge.
Strange, J. M. (2003). '"Tho' lost to sight, to memory dear": Pragmatism, Sentimentality and Working-Class Attitudes towards the Grave, *c*.1875–1914', *Mortality*, 8/2: 144–59.
Strange, J. M. (2005). *Death, Grief and Poverty in Britain*. Cambridge: Cambridge University Press.
Strathern, M. (2005). *Kinship, Law and the Unexpected: Relatives Are Always a Surprise*. Cambridge: Cambridge University Press.
Strawser, B. J., and J. McMahan (2013). *Killing by Remote Control: The Ethics of an Unmanned Military*. Oxford and New York: Oxford University Press.
Sturken, M. (1991) 'The Wall, the Screen, and the Image: The Vietnam Veterans Memorial', *Representations*, 35 (Summer), 118–42.
Summers, J. (2010). *British and Commonwealth War Cemeteries*. Oxford: Shire Publications.
Swain, V. A. J. (1970). 'Franco-Prussian War 1870–1871: Voluntary Aid for the Wounded and Sick', *British Medical Journal*, 3/5721: 511–14.
Sylvester, C. (2013). *War as Experience: Contributions from Internatioanl Relations and Femminist Analysis*. London and New York: Routledge.
Tallet, F. (1992). *War and Society in Early-Modern Europe, 1495–1715*. London: Routledge.

Tarlow, S. (1999). *Bereavement and Commemoration: An Archaeology of Mortality*. Oxford: Blackwell.

Taylor, A. J. P. (1980). *The First World War: An Illustrated History*. New York: Perigee.

Taylor, L. J. (1989). 'Introduction: The Uses of Death in Europe', *Anthropological Quarterly*, 62/4: 149–54.

Tenenti, A. (1989). *Il senso della morte e l'amore della vita nel Rinascimento*. Turin: Einaudi.

Thomas, L.-V. (2001). 'L'Homme e la mort', in J. Poirier (ed.), *Histoire de mœrs, volume 2: Modes and modèles*. Paris: Gallimard, 803–68.

Thomas, W. (2001). *The Ethics of Destruction: Norms and Force in International Relations*. Ithaca, NY: Cornell University Press.

Thucydides (1998). *The Peloponnesian War*, trans. S. Lattimore. Indianapolis and Cambridge: Hackett.

Tilly, C. (1992). *Coercion, Capital, and European States, AD 990–1990*. Oxford: Oxford University Press.

Tirman, J. (2011). *The Death of Others: The Fate of Civilians in America's War*. Oxford: Oxford University Press.

Toner, J. H. (1981). 'American Society and the American Way of War: Korea and Beyond', *Parameters*, 11/1: 79–90.

Townshnd, I. R. (2000). *Casualty Aversion: The New Principle of War?* Newport: Naval War College.

Treaty of Peace with Germany (Treaty of Versailles). 1919. 28 June. https://www.census.gov/history/pdf/treaty_of_versailles-112018.pdf.

Trevisan, C. (2001). *Les Fables du deuil. La Grande Guerre: Mort et écriture*. Paris: PUF.

Tripodi, C. (2021). *The Unknown Enemy: Counterinsurgency and the Illusion of Control*. Cambridge: Cambridge University Press.

Trout, S. (2010). *On the Battlefield of Memory: The First World War and American Remembrance, 1919–1941*. Tuscaloosa: University of Alabama Press.

Troyansky, D. G. (1986a). 'Cult of the Cemetery', in P. H. Hutton (ed.), *Historical Dictionary of the Third French Republic, 1870–1940, Volume 1*, Westport: Greenwood Press, 173–4.

Troyansky, D. G. (1986b). 'Cult of the Fallen Soldier', in P. H. Hutton (ed.), *Historical Dictionary of the Third French Republic, 1870–194, Volume 1*. Westport: Greenwood Press, 351–2.

Troyansky, D. G. (1987). 'Monumental Politics: National History and Local Memory in French *Monuments aux Morts* in the Department of the Aisne since 1870', *French Historical Studies*, 15/1: 121–41.

Turner, B. (2013). 'War Losses and Worldview: Re-Viewing the Roman Funerarty Altar at Adamclisi', *American Journal of Philology*, 134/2: 277–304.

Turse, N., H. Moltke, and A. Speri (2018). 'Secret War', *Intercept*, 21 June. https://theintercept.com/2018/06/20/libya-us-drone-strikes/.

Twain, M. (2008). *The Complete Short Stories of Mark Twain*. Stilwell, KS: Digireads.

Ullmann, W. (1967). *The Individual and Society in the Middle Ages*. London: Methuen & Co.

UNAMA. 2012. *Afghanistan Annual Report 2011: Protection of Civilians in Armed Conflict*, Kabul. http://www.ohchr.org/Documents/Countries/AF/UNAMA_Feb2012.pdf.

Underhill-Cady, J. B. (2001). *Death and the Statesman: The Culture and Psychology of US Leaders in War*. New York: Palgrave.

US Department of Defense (2000). *Kosovo/Operation Allied Force After-Action Report*, Report to Congress, 31 January, Washington, https://webharvest.gov/peth04/20041027022740/www.defenselink.mil/pubs/kaar02072000.pdf.

US Department of the Army (2004). *FMI 3-07.22: Counterinsurgency Operations*. Washington, https://irp.fas.org/doddir/army/fmi3-07-22.pdf.
US Department of the Army (2006). *FM 3-24: Counterinsurgency*. Washington, https://www.hsdl.org/?abstract&did=468442.
Valentino, B. A. (2004). *Final Solutions: Mass Killing and Genocide in the Twentieth Century*, Ithaca, NY: Cornell University Press.
Valentino, B. A., P. Huth, and D. Balch-Lindsay (2004). '"Draining the Sea": Mass Killing and Guerrilla Warfare', *International Organization*, 58/2: 375–407.
Varley, K. (2008). *Under the Shadow of Defeat: The War of 1870–71 in French Memory*. New York: Palgrave Macmillan.
Vennesson, P. (2011). 'War without the People', in H. Strachan and S. Scheipers (eds), *The Changing Character of War*. Oxford: Oxford University Press, 241–58..
Verdery, K. (1999). *The Political Lives of Dead Bodies: Reburial and Postsocialist Change*. New York: Columbia University Press.
Ville de Goyet, C. de (2004). 'Epidemics Caused by Dead Bodies: A Disaster Myth that does not Want to Die', *Pan American Journal of Public Health*, 15/5: 297–9.
Voelz, G. J. (2018). *The Rise of iWar: Identity, Information, and the Individualization of Modern War*. New York: Skyhorse.
Vovelle, M. (1983). *La Mort et l'Occident de 1300 à nos jours*. Paris: Gallimard.
Vucetic, S. (2011). *The Anglosphere: A Genealogy of a Racialized Identity in International Relations*. Stanford: Stanford University Press.
Wagner, S. (2013). 'The Making and Unmaking of an Unknown Soldier', *Social Studies of Science*, 43/5: 631–56.
Wagner, S. (2015). 'A Curious Trade: The Recovery and Repatriation of Vietnam MIAs', *Comparative Studies in Society and History*, 57/1: 161–90.
Wagner-Pacifici, R., and B. Schwartz. (1991). 'The Vietnam Veterans Memorial: Commemorating a Difficult Past', *American Journal of Sociology*, 97/2: 376–420.
Walter, T. (2012). 'Why Different Countries Manage Death Differently: A Comparative Analysis of Modern Urban Societies', *British Journal of Sociology*, 63/1:. 123–45.
Walter, T. (2020). *Death in the Modern World*. Los Angeles: Sage.
Walzer, M. (2006.) *Just and Unjust Wars: A Moral Argument with Historical Illustrations*. New York: Basic Books.
Wasinski, C. (2008). '"Post-Heroic Warfare" and Ghosts: The Social Control of Dead American Soldiers in Iraq', *International Political Sociology*, 2/2: 113–27.
Wasinski, C. (2011). 'On Making War Possible: Soldiers, Strategy, and Military Grand Narrative', *Security Dialogue*, 42/1: 57–76.
Watson, A., and P. Porter (2008). 'Bereaved and Aggrieved: Combat Motivation and the Ideology of Sacrifice in the First World War', *Historical Research*, 83/219: 146–64.
Weber, M. (1949). '"Objectivity" in Social Science', in E. A. Shils and H. A. Finch (eds), *The Methodology of Social Sciences*. New York: Free Press, 72–112.
Weigley, R. F. (1973). *The American Way of War: A History of United States Military Strategy and Policy*. New York and London: Macmillan.
Weiss, P. (1970). *The Persecution and Assassination of Jean-Paul Marat as Performed by the Inmates of the Asylum of Charenton under the Direction of the Marquis de Sade*. London: Calder and Boyars.
Wellerstein, A. (2020). 'Counting the Dead at Hiroshima and Nagasaki', *Bulletin of the Atomic Scientist*, 4 August, https://thebulletin.org/2020/08/counting-the-dead-at-hiroshima-and-nagasaki/.

Wells, R. V. (2000). *Facing the 'King of Terrors': Death and Society in an American Community, 1750–1990*. New York: Cambridge University Press.

Welsh, J. M. (2019). 'The Individualisation of War: Defining a Research Programme', *Annals of the Fondazione Luigi Einaudi: An Interdisciplinary Journal of Economics, History and Political Science*, 53/1: 9–28.

Wendt, A. (1999). *Social Theory of International Politics*. Cambridge: Cambridge University Press.

Wesseling, H. L. (2013.) *The European Colonial Empires, 1815–1919*. London and New York: Routledge.

Wessely, S. (2006). 'Twentieth-Century Theories on Combat Motivation and Breakdown', *Journal of Contemporary History*, 4/2: 269–86.

Whalen, B. E. (2020). 'Political Theology and the Metamorphoses of *The King's Two Bodies*', *American Historical Review*, 125/1: 132–45.

Wilcox, L. B. (2015). *Bodies of Violence: Theorizing Embodied Subjects in International Relations*. New York: Oxford University Press.

Wilkinson, A. (1997). 'Changing English Attitudes to Death in the First World War', in P.C. Jupp and G. Howarth (eds), *The Changing Face of Death: Historical Accounts of Death and Disposal*. London: Macmillan Press, 149–63.

Williams, D. (2009). *Media, Memory, and the First World War*. Montreal: McGill University Press.

Wills, G. (2006). *Lincoln at Gettysburg: The Words that Remade America*. New York: Simon & Schuster.

Wilson, P. (1999). 'Warfare in the Old Regime 1648–1789', in J. Black (ed.), *European Warfare 1453–1815*. London: Macmillan, 69–95.

Wilson, R. (2008). 'Strange Hells: A New Approach on the Western Front', *Historical Research*, 81/211: 150–66.

Wimmer, A. (2014). 'War', *Annual Review of Sociology*, 40: 173–97.

Winter, D. (1978). *Death's Men: Soldiers of the Great War*. London: Penguin.

Winter, J. (1995). *Sites of Memory, Sites of Mourning: The Great War in European Cultural History*. Cambridge: Cambridge University Press.

Winter, J. (2006). *Remembering War: The Great War between Memory and History in the Twentieth Century*. New Haven: Yale University Press.

Winter, J. (2013). 'Introduction', in P. Fussell (ed.), *The Great War and Modern Memory*. Oxford and New York: Oxford University Press, pp. ix–xiv.

Winter, J., and E. Sivan (1999). 'Setting the Framework', in J. Winter and E. Sivan (eds), *War and Remembrance in the Twentieth Century*. Cambridge, Cambridge University Press, 6–39.

Wittman, L. (2011). *The Tomb of the Unknown Soldier, Modern Mourning, and the Reinvention of the Mystical Body*. Toronto: University of Toronto Press.

Woloch, I. (1979). *The French Veteran from the Revolution to the Restoration*. Chapel Hill: University of North Carolina Press.

Woloch, I. (1982). *Eighteenth-Century Europe: Tradition and Progress, 1715–1789*. New York: W. W. Norton & Company.

Wong, L. (2005). 'Leave No Man Behind: Recovering America's Fallen Warriors', *Armed Forces and Society*, 31/4: 599–622.

Wood, M. (2000). *Blind Memory: Visual Representations of Slavery in England and America 1780–1865*. New York: Routledge.

Wood, W. R., and J. B. Williamson. (2003). 'Historical Changes in the Meaning of Death in the Western Tradition', in C. D. Bryant (ed.), *Handbook of Death & Dying, Volume One: The Presence of Death*. Thousand Oaks, CA: Sage, 14–23.

Yarrington, A. (1983). 'Nelson the Citizen Hero: State and Public Patronage of Monumental Sculpture, 1805–18', *Art History*, 6/3: 315–29.

Zambernardi, L. (2010). 'Counterinsurgency Impossible Trilemma', *Washington Quarterly*, 33/3: 21–34.

Zambernardi, L. (2016a). 'Collateral Damage', in Paul Joseph (ed.), *The Sage Encyclopedia of War: Social Science Perspectives*. Thousand Oaks, CA: Sage, 331–3.

Zambernardi, L. (2016b). 'Politics is Too Important to Be Left to Political Scientists: A Critique of the Theory-Policy Nexus in International Relations', *European Journal of International Relations*, 22/1: 3–23.

Zambernardi, L. (2020a). 'The Limits of Power: Knowledge, Ethics, and Foreign Policy in Hans J. Morgenthau's International Theory', *International Relations* (June), 1–20, doi:10.1177/0047117820935621.

Zambernardi, L. (2020b). 'Western Attitudes toward Soldiers' Death: From the Early Modern Period to the Present', *Mortality*, 1–17, doi: 10.1080/13576275.2020.1864723.

Zehfuss, M. (2009). 'Hierarchies of Grief and the Possibility of War: Remembering UK Fatalities in Iraq', *Millennium: Journal of International Studies*, 38/2: 419–40.

Zelizer, V. A. (1994). *Pricing the Priceless Child: The Changing Social Value of Children*. Princeton: Princeton University Press.

Index